Organisms for Genetics

Schools Council
Educational Use of Living Organisms
General Editor: P J Kelly

Organisms for Genetics

Author: L C Comber

HODDER AND STOUGHTON
LONDON SYDNEY AUCKLAND TORONTO

Schools Council Educational Use of Living Organisms Project
Director: P.J. Kelly
Research Fellow: J.D. Wray
This project was established at the Centre for Science Education, Chelsea College, in 1969. Its main aims have been to determine the needs of schools with respect to living organisms, to evaluate the usefulness of various kinds of organisms for educational purposes, and to devise maintenance techniques and teaching procedures for the effective use of appropriate species.

ISBN 0 340 17051 4

First published 1976

Printed in Great Britain for
Hodder and Stoughton Educational,
a division of Hodder and Stoughton Limited, London
Computer Typesetting by Print Origination,
Merseyside, L20 6NS

Contents

Preface

The use of living organisms in schools, while not new, has received considerable emphasis in recent developments in the teaching of biology, environmental studies and allied subjects in colleges and secondary schools, and in many aspects of primary school work. It is a change reflecting the much wider movement in educational thinking which acknowledges both the interest and delight that young people can gain from animals and plants and the importance of fostering an appreciation of the scientific, social, aesthetic and moral issues involved in the study of life and the natural environment. It is a change, also, that presents some very real—but not insurmountable—practical problems for schools.

The implications of using living organisms for education are basically threefold. There needs to be an adequate supply of appropriate organisms and they should be kept healthy and be able to live as naturally as possible. Accommodation for the organisms should allow them to be observed and studied easily but with respect. Maintenance of the organisms should not place an undue burden on teachers and technicians. The Educational Use of Living Organisms publications are intended to assist teachers (including, of course, student teachers), technicians and administrators to contend with these implications.

The books in the series deal with the principles which underlie the effective educational use of living things. They also provide information to help teachers integrate work with organisms into their courses and to cope with the practical, day-to-day problems involved. For teachers and administrators there are technical details of value for planning facilities, and annotated bibliographies provide the guidelines for more detailed studies if they are required. In addition, posters and slide transparencies for use with pupils have been produced complementary to the books.

The series has been produced as part of the work of the Educational Use of Living Organisms project. The project was initiated by the Institute of Biology and Royal Society Biological Education Committee and was established as a Schools Council Project at the Centre for Science Education of Chelsea College (University of London) in 1969. While the project's major financial support was from the Schools Council, the Nuffield Foundation and Harris Biological Supplies have also given generous contributions.

Many people have assisted in a personal capacity and we have particular regard to the interest shown by Mr J A Barker, Mr D J B Copp, Mr T A M Gerrard, Mr B J F Haller, Mr O J E Pullen and Dr C H Selby.

Mr J D Wray undertook much of the research into the project and provided much of the information on which the books are based. In this, he was admirably supported by Mr J B Green as technician and Miss M Hoy and later Mrs E Barton as the project's secretary.

The authors of the books have very kindly been most tolerant of their editor. I would like to express my gratitude to them and our publisher for both their help and forbearance.

This particular book is the product of many years of research and experience in the teaching of genetics by L. C. Comber. It is a delight to acknowledge this and, also, to have the opportunity of thanking him for the many other valuable contributions he has made to biology teaching.

P J Kelly
Director, Educational Use of Living
Organisms Project

1 The use of living organisms in the teaching of genetics in schools

The purpose of this book is to bring together in a compact form, information about the maintenance and use of plants and animals that have been found suitable under ordinary school conditions for the teaching of elementary genetics. It is intended to help teachers select from the wealth of materials available to them those organisms most suited to their needs.

Genetics provides a theoretical structure to modern experimental and analytical biology. Modern biology teaching should be concerned therefore with the systematic erection of this theoretical framework in such a way that pupils not only understand its nature and use, but also appreciate the contributions of the pioneers in this field and have practice in the methods that have proved successful in scientific enquiry. Learning genetics can be a valuable part of any scientific education.

Two important ideas of genetics that should emerge from a school course are, firstly, that the same fundamental materials and processes operate right across the spectrum of life from the lowliest viruses and bacteria to man himself and, secondly, that in the development of a scientific theory concepts are constantly modified as new discoveries make adjustment necessary. Conceived in this way a study of genetics cannot be hurried. Genetical experiments, especially with plants, take time, often measured in years rather than weeks or months.

There are those who say that the interest of young people cannot be sustained over such long periods of time, that they need quick results. This may be so and difficulties on that score are sure to arise but the essential point is that such prolonged studies make for good science education and the skilful teacher will find ways of overcoming the difficulties. Fortunately the subject of inheritance is intrinsically interesting even to quite young children so that an early start can be made. Added to this genetics lends itself to real and satisfying enquiries at a comparatively simple level so that a high degree of personal involvement can arise in a large proportion of pupils.

The pattern of this book is therefore based on two assumptions.

a) That a course in genetics will span at least several years and will consist of a series of related studies each appropriate to the stage it occupies.
b) That a course is experimental and progressive, each step forward in theory being based on the pupil's direct experience.

Although the treatment takes this somewhat idealistic form it does not necessarily render the book any less useful for the reader who wants to cull information for a different purpose. The information is there and the use to which it is put is the responsibility and privilege of the teacher.

The list of organisms dealt with in the book does not attempt to be exhaustive. Two criteria have been used in selection. First, that an organism has been found easy and satisfactory to use under school conditions. Second, that it illustrates some important point of management or teaching use. Some organisms are treated fairly fully as illustrative of general methods and purposes. Others are treated less fully, not necessarily because they are less desirable or important but because it would mean duplicating information. For these only specific details are given.

There remain those organisms which although of undoubted value in research and advanced laboratories have been omitted altogether from the book. A word of explanation about four of the most well-known of them is needed. *Serratia marcescens* which has been used for many years, especially to demonstrate mutation, is now known to be a human pathogen. *Neurospora crassa* has played an important part in the development of genetical theory but under school conditions contamination of culture plates by liberated airborne spores has proved difficult to control. *Arabidopsis thaliana* has many things in its favour as a flowering plant easily cultured under standard conditions but its manipulation in crossing is not easy and school pupils, even at sixth form level, have difficulty with it. *Mormoniella* spp. seemed a most attractive organism when it was first made commercially available in about 1960, especially as it involved a parasite/host relationship. However, it has not proved successful for school work and it is not now easily obtained. The technical difficulties encountered with bacteriophages have led to their exclusion also.

The book should be looked upon therefore as an essentially practical guide for teaching genetics in schools. The references should enable the more experienced teacher to explore an even wider range of organisms and to go deeper into the experimental evidence for modern genetical theory.

The 'outline scheme' (Fig 1) sets out, in the form of a flow diagram, the development of genetical theory suitable for secondary level. It also shows the points

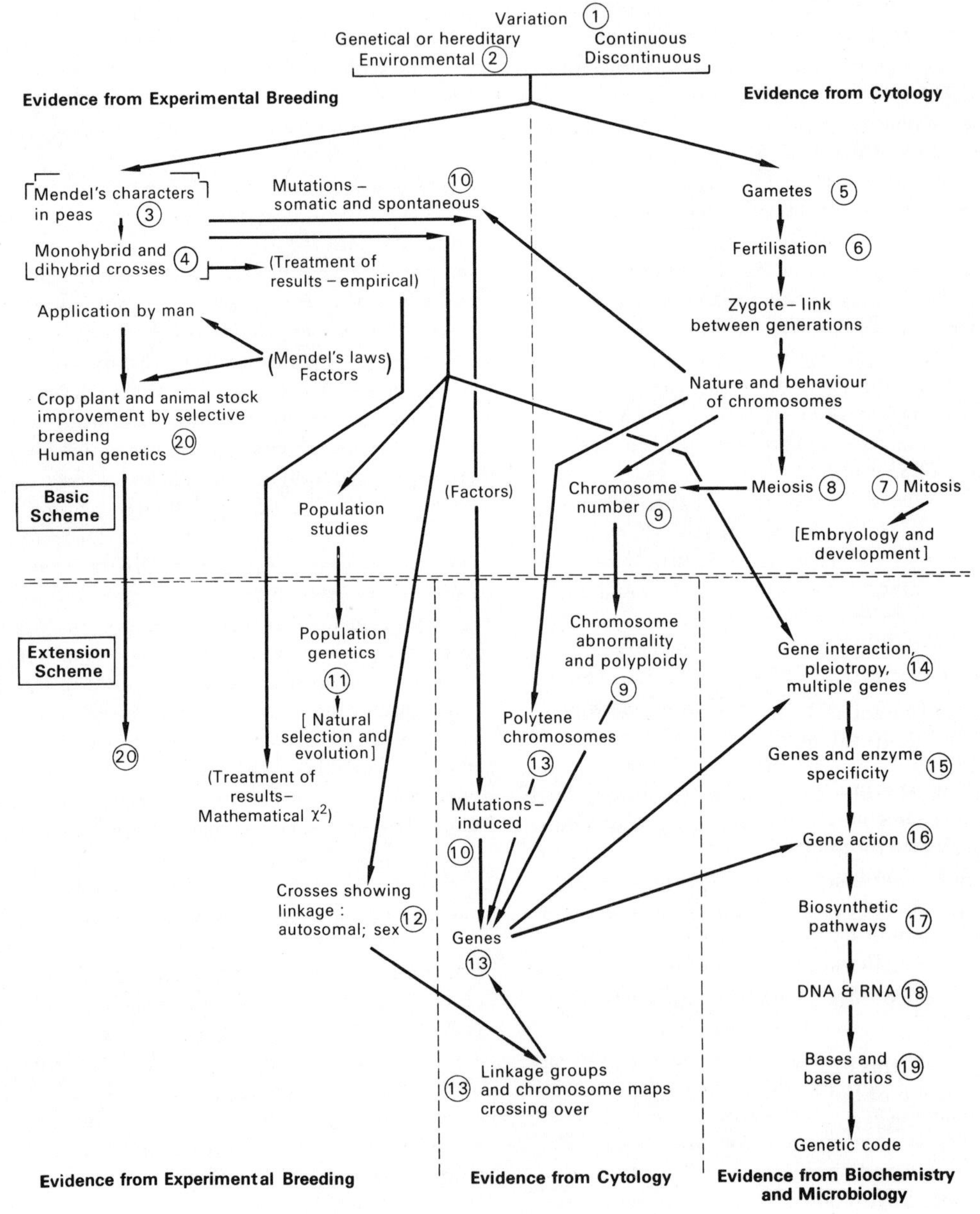

Figure 1 The 'Outline Scheme'

at which the work in genetics links with other major aspects of biology. The scheme attempts to draw upon and relate three lines of evidence and to provide demonstrations for major steps in the argument beginning from the experience and recording of biological variation appropriate to the lowest forms in secondary schools and leading to some understanding of the genetic code and its implications in the final school years.

The circled numbers in the scheme refer to the practical activities or investigations associated with the various steps. These are summarised in Table 1.

Table 1 Summary of work indicated in the outline scheme

1 **Field, garden and laboratory studies on variation**
Effects upon variation of in-breeding and out-breeding mechanisms and comparative studies e.g. wheat and rye. Human characteristics, e.g. blood groups.

2 **Environmental variation**
Clones grown under different environmental conditions e.g. *Bryophyllum* or dandelion in different soils, illumination or water status. Xantha barley in dark and light. Bar eye *Drosophila* at different temperatures. Red canary with and without carotene in diet. Himalayan rabbit at high and low temperatures.

3 **Mendel's characters in the organism which he originally used: Edible pea**

4 **Breeding experiments; monohybrid and dihybrid crosses; Mendelian ratios**
Mouse (coat colour and other characteristics), *Drosophila* (body, eye and wing characters), maize (endosperm food store and aleurone colour), tomato (seedling characters) *Antirrhinum* (flower pigments, disease resistance and plant height).

5 **Gametes**
Eggs and sperms of *Pomatoceros* and *Fucus*. Examination of bull semen. Growth of pollen tubes (see references 5, 18 and 19).

6 **Fertilisation and development**
Rhabditis, frog, toad, hen. The link here with embryology is so evident that no attention is given to these topics in the text. See references 5, 17, 18 and 19.

7 **Mitosis**
Squash preparations of broad bean, onion and *Tradescantia* root tips.

8 **Meiosis**
Squash preparations of spermatogenesis in grasshopper and locust and of pollen mother cells in *Tradescantia* and other plants.

9 **Chromosome numbers and polyploidy**
Crocus, *Drosophila* from cytological preparations and from linkage groups. Polyploidy in *Primula kewensis*, *Senecio cambriensis*, x *Geum intermedium* etc.

10 **Mutations**
Somatic: spontaneous in *Pelargonium* spp. etc. Germinal: spontaneous and induced in yeast by ultra-violet light. Germinal: induced in barley by gamma radiation and mutagen treatment.

11 **Population studies and population genetics**
Drosophila, *Tribolium*, *Senecio* (rayed and rayless). Bead models (not referred to in text; see reference 20).

12 **Breeding experiments showing linkage, both autosomal and sex**
Drosophila, tomato, mouse, poultry.

13 **Final establishment of the gene from polytene chromosomes, linkage groups and chromosome maps**
Chromosome studies in *Drosophila*. Polytene chromosomes in *Chironomus*, *Simulium* or broad bean and study of crossing over in *Sordaria*.

14 **Gene interaction, pleiotropy and multiple genes**
Breeding experiments with mouse, pea, maize, *Antirrhinum*. Dwarfing genes in wheat.

15 **Genes and enzyme specificity**
Cyanogenesis in *Trifolium*. Three levels of expression of a gene in peas.

16 **Gene action**
Antirrhinum flower pigments. Response of dwarf maize to gibberellic acid and IAA. Response of thiamineless tomato to thiamine.
Complementation in yeast or *Coprinus*.

17 Biosynthetic pathways
Coprinus (ethanolamine-choline) and tomato (pyrimidine-thiamine).

18 Identification of DNA and RNA
Feulgen and methyl-green-pyronine methods in root tip cells of peas and beans. Extraction of DNA and RNA from pea root tips.

19 Hydrolysis of commercial DNA; separation and identification of products by paper chromotography and ultra-violet light.

20 Applications of genetics; crop plant and animal stock improvement by selective breeding
Wheat pedigrees in plant breeding. Human genetics.

Table 2 is a list of the organisms considered in the book arranged in order of chapters. The list indicates, again by the circled numbers shown in the 'outline scheme', the uses to which the given organism can be put in the teaching of genetics and its other uses in a school biology course. An indication is also given of the suitability of the organisms and an assessment of the problems related to their supply, maintenance and use.

Table 2 Organisms—uses, suitability and assessment (See page 17 for key)

1 Organisms	2 Uses in genetics and role in the outline scheme	3 Other uses	4	5	6	7	8	9	10
1 **Vertebrate Animals —Mammals**				Rating on 3 point scale					
1.1 *Mouse *Mus musculus*	Variation. Mendelian ratios. Linkage. Quantitative inheritance Gene interaction. *1, 4, 12, 14*	Characteristics and life history of mammals. Behaviour. Reproduction.	B	1	2	1	2	2	1
1.2 Rabbit *Oryctolagus cuniculus*	Mendelian ratios. Temperature sensitive gene action. *4, 16*	Characteristics of herbivorous mammal. Mammalian anatomy (dissection). Reproduction.	B	1	1	1	3	2	2
1.3 Guinea pig *Cavia porcellus*	Mendelian ratios. *4*	Reproduction.	B	1	1	1	3	2	1
1.4 Syrian hamster *Mesocricetus auratus*	Mendelian ratios. *4*	Reproduction.	B	1	1	1	2	2	2
1.5 Laboratory rat *Rattus norvegicus*	Mendelian ratios. *4*	Mammalian anatomy (dissection). Behaviour.	B	1	3	2	2	2	1
1.6 Mongolian gerbil *Meriones unguiculatus*	Mendelian ratios. *4*	Behaviour. Reproduction	B	2	1	1	2	2	1
1.7 Man *Homo sapiens*	Variation and inheritance generally	Human biology	B						

1 Organisms	2 Uses in genetics and role in the outline scheme	3 Other uses	4	5	6	7	8	9	10
1.8 Cat *Felis catus*	Variation and inheritance generally		B						
1.9 Cattle *Bos taurus*	Variation and inheritance generally		B						
1.10 Sheep *Ovis aries*	Variation and inheritance generally		B						
1.11 Pig *Sus scrofa*	Variation and inheritance generally		B						
1.12 Horse *Equus caballus*	Variation and inheritance generally		B						
2 Other Vertebrate Animals									
2.1 Guppy *Poecilia reticulata*	Mendelian ratios, Sex linkage. *4, 12*	Fish locomotion. Life history. Behaviour.	B	2	1	1	2	1	2
2.2 Budgerigar *Melopsittacus undulatus*	Variation. Mendelian ratios. Linkage. *1, 4, 12*	Bird flight, life history. Colouration.	B	1	1	1	2	2	1
2.3 Canary *Serinus canarius*	Mendelian ratios. Hybridisation. Variation environmental/genetical interaction. *2, 4, 16*	Bird biology.	B	1	1	1	2	1	1
2.4 *Fowl *Gallus domesticus*	Mendelian ratios modified. Gene interaction. Sex linkage. Animal breeding. *4, 12, 14, 20*	Evolution. Embryology.	B	2	1	1	2	3	2
3 Invertebrate Animals									
3.1 **Drosophila melanogaster*	Variation. Mendelian ratios. Linkage: Temperature and gene action. Polytene chromosomes. Population genetics. *1, 2, 4, 11, 12, 13*	Life history. Behaviour. Taxes.	B	1	2	2	1	2	2
3.2 **Tribolium confusum*	Mendelian ratios. Population genetics. *4, 11*	Life history. Behaviour. Pest of stored products.	B	1	3	3	1	2	1

1 Organisms	2 Uses in genetics and role in the outline scheme	3 Other uses	4	5	6	7	8	9	10
3.3 **Chorthippus spp.*	Cytology. *8*	Life histories.	E	2/3	2	2	1	2	3
3.4 Locusts *Schistocerea gregaria*	Cytology. *8*	Behaviour. Insect type for anatomy and physiology.	B	1	2	2	1	2	2
3.5 *Locusta migratoria*	Cytology. *8*								
4 **Flowering plants—Greenhouse Plants**									
4.1 *Tomato *Lycopersicon esculentum*	Variation. Mendelian ratios. Gene interaction. Plant breeding. Biosynthetic pathways *1, 2, 4, 12, 13, 14, 17*	Succulent berry.	B	1	2	1	1	2	2
4.2 Geraniums *Pelargonium spp.*	Mendelian ratios modified. Polyploidy. Somatic mutation. Chimaeras. Variation. *1, 4, 9, 10*	Physiological demonstrations and investigations. Vegetative propagation. Dispersal. Decoration.	B	1	1	1	1	2	2
4.3 Primulas *Primula kewensis,*	Polyploidy *9*	Decoration. Taxonomy.	E	2	1	1	2	2	2
P. sinensis,	Mendelian ratios. *4*	Decoration. Taxonomy.							
P. obconica,	Mendelian ratios. *4*	Decoration. Taxonomy.							
P. malacoides	Mendelian ratios. *4*	Decoration. Taxonomy.							
4.4 *Rhoeo discolor*	Cytology. *8*	Decoration.	E	2	2	2	2	1	2
4.5 Cucumber *Cucumis sativus*	Polyploidy. Mendelian ratio (physiological character). *4, 9*	Plant anatomy.	E	1	2	1	3	2	2
5 **Flowering Plants—Half-hardy Annuals**									
5.1 *Snapdragon *Antirrhinum majus*	Variation. Complex Mendelian ratios. Unstable genes. Polyploidy. Plant breeding. *1, 2, 4, 10, 14, 20*	Life cycle of a flowering plant. Pollination mechanisms and investigations. Dispersal. Fungal diseases.	B/E	1	1	1	1	2	1
5.2 *Maize *Zea mays*	Monohybrid and dihybrid ratios. Epistasy. Segregation in pollen. Gene action. *4, 14, 16*	Fruit structure and germination. Floral mechanisms. Growth of pollen tubes. Mechanical structures. Plant morphology and anatomy. Productivity. Evolution.	B/E	1	2	1	2	3	2

1 Organisms	2 Uses in genetics and role in the outline scheme	3 Other uses	4	5	6	7	8	9	10
5.3 Stock *Matthiola incana*	Variation, Mendelian ratios. Plant breeding and commercial applications. *1, 4, 20*	Decoration	E	1	1	2	1	2	1
5.4 *Ipomea purpurea*	Mendelian ratios *4*	Decoration	E	2	1	2	2	1	2
5.5 *Mirabilis jalapa*	Mendelian ratios *4*	Decoration	E	2	1	2	2	1	1
5.6 Tobacco *Nicotiana affinis*	Mendelian ratios *4*	Decoration	E	1	1	2	2	1	1
5.7 Petunia *Petunia hybrida*	Mendelian ratios *4*	Decoration	E	1	1	2	2	1	1
6 Flowering Plants— Hardy field and garden plants. Annuals and Biennials									
6.1 *Barley *Hordeum spp.*	Variation. Environmental and genetical interaction. Induction of mutations. Lethal genes and their persistence in a population. Nature of a species. *1, 2, 4, 10*	Inbreeding and its application in an economic crop. Malting and its biological implications. Plant evolution and history.	E	2	2	3	2	2	1
6.2 Wheat *Triticum spp.* 6.3 Rye *Secale cereale*	Inbreeding and outbreeding (wheat and rye compared). Continuous variation and its explanation in terms of multiple genes. Polyploidy. Plant breeding. *1, 9, 20*	Economic crop showing adaptive capabilities. Evolution and history.	E	2	2	3	2	2	1
6.4 *Edible Pea *Pisum sativum*	Demonstration of the characters Mendel used. Monohybrid and dihybrid ratios. Pleiotropy. Three levels of expression of a gene. Extraction of DNA and RNA. *3, 4, 16, 18*	Seed germination. Growth studies. Plant morphology and physiology. Root nodules. Floral mechanisms.	B	1	2	2	2	2	2
6.5 Broad Bean *Vicia faba*	Root tip squashes 7, *13*	Germination and seed structure. Plant anatomy and physiology.	B	1	2	2	1	1	1

1 Organisms	2 Uses in genetics and role in the outline scheme	3 Other uses	4	5	6	7	8	9	10
6.6 Sweet Pea *Lathyrus odoratus*	Monohybrid and dihybrid ratios. Gene action. *4, 16, 20*		B	1	1	2	2	2	2
6.7 Groundsel and Ragwort *Senecio spp.*	Monohybrid ratio. Population genetics. Polyploidy. *4, 9, 11*	Ecology of weeds.	E	2	2	3	1	2	2
6.8 Beets *Beta spp.*	Mutation and selection *10, 20*	Evolution under domestication.	B	1	1	3	1	2	2
6.9 Brassicas *Brassica spp.*	Variation. Polyploidy. *1, 9, 20*	Evolution under domestication.	B	1	2	1	1	3	2
6.10 Radish *Raphanus sativa*	Variation. Dihybrid ratio. *1, 4, 20*	Evolution under domestication	B	1	2	1	1	1	2
6.11 Wallflower *Cheiranthus cheiri*	Monohybrid ratio. Gene action. *4, 16, 20*	Floral structure.	B	1	1	2	1	2	2
7 **Flowering Plants—Hardy field and garden plants Perennials Ferns**									
7.1 *Roses *Rosa spp.*	Mutation. Hybridisation. *10*	Decoration	B	1	1	1	1	3	2
7.2 Hazel *Corylus avellana*	Mutation. *10*								
7.3 Willows *Salix spp.*	Mutation. *10*								
7.4 Juniper *Juniperus communis*	Mutation. *10*								
7.5 Dyer's greenweed *Genista tinctoris*	Mutation. *10*								
7.6 *Kerria japonica*	Mutation. *10*								
7.7 Ling *Culluna vulgaris*	Variation. Mutation. *1, 10*								
7.8 Mountain Ash *Sorbus aucuparia*	Mutation. *10*								
7.9 Primroses *Primula spp.*	Hybridisation. Polyploidy. *4, 9*								

1 Organisms	2 Uses in genetics and role in the outline scheme	3 Other uses	4	5	6	7	8	9	10
7.10 Campions *Melandrium spp.*	Monohybrid ratio. Population genetics. *4, 9, 11*								
7.11 Wood avens *Geum spp.*	Polyploidy. *9*								
7.12 Docks *Rumex spp.*	Polyploidy. *9*								
7.13 Rushes *Juncus spp.*	Polyploidy. *9*								
7.14 *Fatshedera* x *Lizea*	Intergeneric hybrid. Polyploidy. *9*								
7.15 *Gaulnettya* x *wisleyensis*	Intergeneric hybrid. Polyploidy. *9*								
7.16 Foxgloves *Digitalis spp.*	Interspecific hybrid. Polyploidy. *9*								
7.17 Blackberry *Rubus spp.*	Interspecific hybrid. Polyploidy. *9*								
7.18 Valerian *Valeriana officinalis*	Polyploidy. *9*								
7.19 Campions *Silene spp.*	Polyploidy. *9*								
7.20 Lesser Celandine *Ranunculus ficaria*	Polyploidy. *9*								
7.21 Polypody *Polypodium spp.*	Polyploidy. *9*								
7.22 *Tradescantia spp.*	Polyploidy. Cytological evidence—meiosis and mitosis. *7, 8, 9*		B	1	2	2	1	1	2
7.23 *Laburnocytisus*	Graft hybrid. Chimaera. *9*								
7.24 Onions *Allium spp.*	Cytology. *7*	Vegetative propagation.	B	1	2	2	1	1	1
7.25 Crocus *Crocus balansae*	Cytology. *7*	Vegetative propagation.	B	1	2	2	1	1	1
7.26 Bluebell *Endymion non-scripta*	Cytology. *7, 8*	Vegetative propagation.	B	1	2	2	1	1	1
7.27 Hyacinth *Hyacinthus spp.*	Cytology. *7, 8*	Vegetative propagation.	B	1	1	2	1	1	1

1 Organisms	2 Uses in genetics and role in the outline scheme	3 Other uses	4	5	6	7	8	9	10
7.28 Red hot poker *Kniphofia spp.*	Cytology. *8*	Vegetative propagation.	B						
7.29 Peony *Paeonia spp.*	Cytology. *8*	Vegetative propagation.	B	1	1	2	1	1	1
7.30 *White clover *Trifolium repens*	Variation. Competition. Gene Action (Cyanogenesis). *1, 2, 15, 16*	Vegetative propagation.	E	2	2	2	1	2	2
8 Micro-organisms—Fungi and Bacteria									
8.1 *Yeasts *Saccharomyces spp*	Life cycles. Mating types. Nutritional and respiratory mutants. Complementation. Induction of mutations by ultra-violet radiation. *10, 16*	Microbiological techniques. Anaerobic respiration. Fermentation. Economic applications.	B	1	2	2	1	3	2
8.2 **Sordaria fimicola*	Linear asci. Crossing over. Chromosome mapping. Temperature effect on rate of crossing over. *13*	Fungal life cycles.	E	1	2	3	3	1	2
8.3 **Coprinus lagopus*	Life cycles. Mating types. Auxotrophs. Segregation and linkage in haploid progenies. Complementation. Biochemical pathways. *4, 10, 16, 17*	Natural history of a toadstool. Ecology.	E	1	2	3	3	1	3
8.4 *Aspergillus niger*	Environmental and genetical interaction and its economic application. *2*	Fungal life cycles.	E	1	2	3	3	1	2
8.5 *Fusarium culmorum*	Environmental variation due to the availability of nutrients. *2*	Fungal disease.	E	2	3	2	3	1	2
8.6 *Bacillus subtilis*	Environmental variation due to the availability of oxygen. *2*	Bacterial action.	E	1	2	2	3	1	2
8.7 *Sarcina lutea*			E	1	2	2	3	1	2

Organisms—uses, suitability and assessment

Those marked with an asterisk are considered in some detail and may be regarded as 'key' organisms.

Uses (Columns 2 and 3)

The uses of the organisms in genetics and their role in the 'outline scheme' is given in column 2. Other possible uses are given in column 3.

Suitability (Column 4)

The outline scheme has been divided into two parts—basic (B) and extension (E). The basic scheme is suitable for an introductory course in genetics from an elementary beginning up to CSE or GCE 'O' Level. The extension scheme is an appropriate addition to this for a more advanced course leading to GCE 'A' Level and beyond.

Assessment (Columns 5 to 10)

An assessment of the problems associated with supply, student reaction and maintenance is given, for those organisms commonly kept or used in schools, under the following headings:

Current availability (Column 5)

Are the genetic variants of the organisms readily available either in the wild, or from suppliers?
Rating: 1 Readily available; 2 Some difficulty; 3 Difficult to obtain.
Organisms in need of conservation, possibly at a local level, should never be taken. This applies particularly to *Chorthippus* spp. and some of the bulbous plants.
All organisms should be known as far as is possible to be free from disease and pests. Small mammals must be obtained from reputable suppliers (see pages 112 to 113, and references 9, 10, 11, 34 and 36).

Student response (Column 6)

Are the organisms attractive to children?
Rating: 1 Attractive; 2 Neutral; 3 Unattractive.

Student manipulation (Column 7)

Can students handle the organism easily and safely? Are the phenotype characters clear and easily seen, preferably without the aid of visual magnification?
Rating: 1 Easy; 2 Somewhat difficult; 3 Difficult

Facilities (Column 8)

What apparatus and equipment is required in maintenance and use?
Rating: 1 Only simple cheap apparatus required; 2 More complex and dearer apparatus and equipment required; 3 Expensive and complex pieces of apparatus and/or equipment necessary.

Space (Column 9)

What space is required to keep the organisms?
Rating: 1 Requires little space (up to 1 m^2); 2 Moderate space required (up to 5 m^2); 3 Large space necessary (more than 5 m^2).

Maintenance (Column 10)

Is the organism simple and not time consuming to maintain in good health and to prepare for use?
Rating: 1 Easy to maintain; 2 Some difficulties in maintenance; 3 Difficult to maintain.

Notation and conventional signs

Symbols for genetical characters and genes are obviously of great value but although some uniformity of practice has been achieved there is still variation.

The following summary explains the most common practices:

1. The wild type gene is shown as +. This is especially convenient when several different pairs of alleles are being represented at the same time. This is standard practice in *Drosophila*, mouse and fungal genetics.
2. Mutants are given suitable symbols consisting of one, two or three descriptive letters. Thus in mice b is the symbol for the recessive brown coat mutant and bt the symbol for belted, which has a belt of lighter colour round its middle.
3. The dominant allele is shown in capital and the recessive in lower case letters, for example, the dominant black gene in the mouse is shown as B and the recessive brown as b.
4. The genotype is indicated by symbols for two characteristics. For example Bb, B/b or $\frac{B}{b}$ indicates the heterozygote black carrying brown mouse; BB, B/B or $\frac{B}{B}$ and bb, b/b or $\frac{b}{b}$ for the homozygous black and brown mice respectively.
5. The genotype is sometimes written as B—, B/— or $\frac{B}{-}$ since the second gene may be B or b and which one it is may not be known (however see 9 below).
6. When there are more than two alleles at a given locus, they are shown by suffixes. Thus in the albino series of mouse mutants C is the symbol for the dominant for full colouration, c^{ch} for chinchilla, c^{e} for extreme dilution and c for absence of colour.

7 When two or more pairs of alleles are involved the genotype is indicated in the form Cc,Bb or CB/cb or $\frac{CB}{cb}$. In this case they indicate a heterozygous black mouse obtained in the F_1 of a cross between a pure black mouse, CC, BB and a pure albino mouse cc, bb. The conventions CB/cb and $\frac{CB}{cb}$ are particularly useful in investigations into linkage since they imply that C and B are on one of the two homologous chromosomes and c and b on the other (see also coupling and repulsion).

8 When a cross is being shown symbolically it is now usual to show the female, or in the case of plants the seed parent, first. For example, CB x cb or CC,BB x cc,bb represents a cross between a pure black female mouse and a pure albino male mouse. At one time for animals the heterogametic sex was usually given first but this practice has almost disappeared.

9 Some confusion still exists over symbols involving the X and Y sex chromosomes. XX obviously represents the homogametic sex and XY and XO the heterogametic sex, depending upon whether a Y chromosome is present or not. Differences arise when the genes in question are situated on the non-homologous section of the X chromosome. An example is the tabby gene Ta in the mouse. The male can be written TaY and the female as TaX but it is better to denote the male as TaX/—Y and the female as TaX/TaX which makes the position quite clear.

Symbols for the succeeding generations of a breeding experiment

Parental generation	P or F_0
First filial generation	F_1
Second filial generation	F_2 and so on.

When mutations are being induced by irradiation or treatment with mutagens the respective generations are denoted by X_1 or M_1, X_2 or M_2 etc.

Signs for the sexes

♀ female	This was the Greek symbol for the goddess Venus and represents a hand mirror which presumably the beautiful goddess was proud to gaze into. The same sign was used by mediaeval astronomers for the planet Venus and by the alchemists for copper on the grounds that the mirror would have been of polished copper.
♂ male	This was the sign for the god Mars and represents the shield and spear of the warlike god. It was also used by astronomers for the planet Mars and the alchemists for iron.
☿ hermaphrodite	

Glossary

Allelomorph (= Allele). One member of a pair of alternative characters. The term is also applied to the alternative hereditary units or genes (q.v.) present in somatic or body cells one having been contributed by the female parent and the other by the male parent.
Multiple allelomorphs (alleles). The term used to describe the condition when there are more than two alternative characters or when more than two alternative genes can occupy the same position or locus (q.v.) on a chromosome.

Auxotroph (ic). A term used in fungal and bacterial genetics to describe a mutant which cannot grow on the minimal medium required by the wild type, and needs some nutritional supplement. See also prototroph.

Autosomes. See chromosomes.

Back cross. See cross.

Bases, nitrogenous. The chemical units or nucleotides, usually the purines adenine and guanine and the pyrimidines thymine and cystosine, that form pairs in DNA. In RNA thymine is replaced by uracil.

Character. A visible or otherwise assessable feature of an organism that arises as the expression of a particular hereditary unit or gene.

Chimaera. A patch of tissue in an organism which is of different genetic constitution from that of the rest of the organism and hence of different appearance. These patches arise as the result of somatic mutations and in some plants that are propagated vegetatively, such as pelargoniums and potatoes, the whole plant may be a chimaera with one layer of tissue genetically different from the other two.

Chromosomes. Deeply staining threadlike structures in the cell nucleus which are composed largely of DNA and which undergo highly organised patterns of change during cell division.

Chromosomes: homologous. The chromosomes, usually two in number, from a normal somatic chromosome complement, that are similar in structure and which come together to form pairs during the zygotene stage of meiosis.

Chromosome: map. A line representing a particular chromosome with the position of the genes known to lie in the chromosome marked on it in the order in which they are believed to occur and spaced at relative distances determined from the frequency of crossing over between them.

Chromosome: number. The number of chromosomes in a normal somatic cell of a species. This is usually written as 2n = 46 (as in man for example)

Diploid number. The same as the chromosome number except in haploid and polyploid species and individuals.

Haploid number. The number of chromosomes in the normal complement of a gamete of a species and of the haploid generation of a species with alternation of generations. It is usually written as n = 23 (as in man for example)

Polyploid number. (triploid, tetraploid, hexaploid etc) The number of chromosomes in the chromosome complement of an organism or species which has more than two sets of homologous chromosomes. This is usually written as 3x =21, 4x =28, 6x =42 etc. (as in roses for example) where x is the number of chromosomes in the basic set or genome.

Chromosomes: polytene. Unusual, many-stranded chromosomes that occur, for example, in the salivary gland cells of the larvae of certain dipterous flies and in the suspensor cells of the embryo of the broad bean. As they are in effect bundles of many homologous chromosomes joined lengthwise studies of their structure gives useful information about the minute structure of single chromosomes.

Autosomes. All those chromosomes in a set (= complement) of chromosomes which are not the X and Y chromosomes (= sex chromosomes).

Chromosomes, sex. Unpaired or incompletely paired chromosomes in a normal chromosome complement, usually one or two in number, which determine the sex of an organism. One sex, the homogametic sex, has normally two fully homologous sex chromosomes indicated as XX while the other sex, the heterogametic sex, has either only one sex chromosome (XO) or two sex chromosomes one of which, the Y chromosome, is different in some way from the X chromosome (XY). The Y chromosome is often shorter than the X chromosome which has therefore homologous and non-homologous sections. Genes on the non-homologous section are responsible for sex linked characters.

Complementation. A term applied in fungal and bacterial genetics to the process by which two recessive mutants in the haploid phase can supply each other's deficiency when they are brought together in the diploid condition.

Cross. The planned mating or pairing of two individuals in such a way as to produce offspring of known parentage.

Cross: back. Usually applied to a cross between a heterozygote and its recessive parent or parental genotype but it may be applied to either parent or parental genotype.

Cross: dihybrid. A cross in which attention is focused upon two pairs of contrasting characters by which the parents differ.

Cross: monohybrid. A cross in which attention is focused on one pair of contrasting characters by which the parents differ.

Cross: reciprocal. A cross which is similar to a particular cross except that the male and female carry the opposite characters.

Cross: self (Selfing). Strictly applied to cases in which the male gametes of an organism, usually a plant, are used to fertilise female gametes of the same organism but it is often applied to the fertilisation of female gametes by organisms with the same genotype with respect to the characters under investigation.

Cross: test. A particular form of back-cross involving linked characters and designed to ascertain the degree of linkage between them.

Crossing-over (Cross-over). Originally applied to the

occurrence of new combinations of linked characters the term is now also used for the exchanges that take place between homologous chromosomes, or more correctly chromatids, during meiosis.

Cross-over frequency. A useful figure determined as the number of recombinants/total number of individuals from a backcross (testcross) of an organism heterozygous for two linked characters with the double recessive (see chromosome map and recombination).

Coupling. The condition in a heterozygote carrying linked characters in which the two dominant or wild type characters are derived from one parent and the two recessive or mutant characters from the other. This condition is represented symbolically as $\frac{AB}{ab}$. See also repulsion.

Cultivar. A cultivated variety of a plant species.

Cytology. The study of plant and animal cells including especially the behaviour of cells during their division.

Dihybrid cross. See cross.

Dihybrid ratio. The theoretical 9:3:3:1 ratio of the numbers of the four different phenotypes obtained from a dihybrid cross in which both characters display dominance.

Dikaryon. A fungal mycelium consisting of cells each of which contains two nuclei derived from different sources and often of two different mating types. See also monokaryon.

Diploid. Applied to a cell or to an individual organism carrying two homologous sets of chromosomes and hence with the diploid (2n) chromosome number (q.v.).

Dominant. The term used by Mendel to describe a character from a pair of contrasting characters being used in a breeding experiment which was expressed in all the individuals in the F_1 generation and in three-quarters of those in the F_2 generation. Such a character appears to have the effect of suppressing the expression of the other character of the pair. See also recessive.

Enzyme. An organic catalyst. Enzymes are usually proteins and they bring about chemical actions, which take place only slowly or not at all in their absence, but are themselves unchanged at the end of the reaction.

Epistasis. The phenomenon in which one gene affects, usually by suppression, the expression of another, non-allelic gene.

Factor. A term used in the early study of genetics to denote the physical element or unit which was carried from one generation to the next in the germ cells and which was the physical basis of the expression of a character. Factors are therefore synonymous with genes.

Fertilisation. The process in sexual reproduction in which the nucleus of the male gamete unites with the nucleus of the female gamete.

Filial generations. The generations of offspring that succeed a genetical cross.

First filial generation F_1. The generation derived immediately from the two parents or parental types

Second filial generation F_2. The generation derived from the random mating of the individuals of the F_1 generation amongst themselves.

Gamete. A germ or sex cell which unites with another germ cell in the act of fertilisation to form a zygote. Gametes usually have the haploid number of chromosomes and the zygote the diploid number.

Gene. The physical element or unit (factor) of heredity. The meaning of the term has been modified, and is still being modified, as the nature and method of functioning of the genetic material has become better understood. The term gene is now correctly applied to the unit of function (the cistron) which may be different from the unit of mutation (the muton) and the unit of recombination (the recon).

Gene: lethal. A gene whose full expression is so severe that it causes death. Most lethal genes are recessive and are carried in populations from one generation to the next in the heterozygous condition.

Gene: major. Defined as a gene capable of mutating in such a way as to permit analysis by Mendelian type breeding experiments.

Gene: multiple. The original term for what are now usually called polygenes.

Polygene. One of a set of genes each of which has only a minor effect upon a quantitative character but whose effects are additive and thus give rise to continuous variation.

Genotype (genotypically). The hereditary or genetical constitution of an individual as distinct from its phenotype (q.v.).

Graft hybrid. A more or less stable plant produced by grafting tissue of one species (the scion) on to part of another species (the stock).

Haploid. A term applied to a cell, tissue or individual with a single set of chromosomes, that is with the haploid number (n) of chromosomes in each cell (see also chromosome number, diploid and polyploid).

Heterogamy (-gamous,-gametic). Having two kinds of sex cells, male and female, determined by the presence of X and Y sex chromosomes. The sex with the constitution XO or XY is known as the heterogametic sex (male in mammals and many insects and female in birds) and the one with the constitution XX as the homogametic sex (female in mammals and many insects, male in birds).

Heterothallic. Refers to species of fungi in which the thallus is self-sterile or incompatible with itself and which depends upon a thallus of another and compatible mating type to bring about sexual union. *Neurospora* is heterothallic.

Homologous chromosomes. Chromosomes or parts of chromosomes that can pair together in a highly specific way with their corresponding portions aligned closely together as happens at the zygotene stage of meiosis.

Homothallic. Refers to species of fungi in which sexual reproduction can occur between parts of the same thallus. That is, the fungus is self-compatible. *Sordaria* is homothallic.

Hybrid. The offspring produced by crossing individuals of two different varieties, species, or even occasionally, genera.

Hybrid vigour or Heterosis. The increased vigour often to be found in hybrids, for example, hybrid corn (maize).

F_1 hybrids. Plants, and more frequently seeds of plants, produced by hybridisation. They have the advantage that they are more uniform in their characteristics than those produced by open pollination.

Hybridisation. The crossing of two different varieties, species or even genera of plants and animals.

Interspecific hybridisation. The crossing of individuals belonging to different species.

Intergeneric hybridisation. The crossing of individuals belonging to different genera.

Interference. The interaction between cross-overs in experiments on linkage which reduces the probability of another cross-over occurring, or, if one does, of its occurring between the same two chromatids.

Linkage. The tendency for certain characters and genes to be associated together. This is now known to be due to the genes involved being situated on the same chromosome. This phenomenon was an important piece of evidence in the development of the chromosome theory of inheritance.

Autosomal linkage. The linkage between genes on autosomes.

Loose linkage. A slight degree of linkage approaching completely independent assortment which is due to the genes being widely separated, though on the same chromosome, and hence with a high probability of crossing over taking place between them.

Sex linkage. The linkage shown between certain characters and the sex of the individuals carrying them so that the characters are inherited in a certain pattern. This is due to the genes concerned being situated on the non-homologous sections of the X chromosome.

Tight or close linkage. Linkage such that the characters involved seldom segregate independently owing to the genes concerned being close together on the same chromosome and hence with a low probability of crossing-over taking place between them.

Locus. The site on a chromosome occupied by a particular gene and its alleles.

Mating type. The condition in a fungal species that determines whether or not a particular thallus can form a sexual union, or in other words is compatible, with another specified thallus. Thus a fungus of mating type A can unite with a thallus of mating type a but not with another of the same mating type A. Mating types are often complex and involve several pairs of genes. Similar compatibility phenomena occur in higher plants also.

Meiosis. The two nuclear divisions with only one division of the chromosomes that have the effect of halving the chromosome number and producing four haploid nuclei that is essential if the chromosome number is to be kept constant in any sexually reproducing species.

Mendel's 'laws'. The results of Mendel's pioneering

experiments on the hybridisation of peas are often stated as two 'laws' though it must be emphasised that Mendel himself never expressed them in this form.

1 Segregation. The hereditary factors (units or genes) that are brought together from the two parents in the process of fertilisation separate, or segregate, unchanged in the later formation of gametes, one from each pair passing into each gamete.

2 Independent assortment. When two or more pairs (alleles) of hereditary factors (units or genes) are brought together by hybridisation all the possible combinations between them occur in the gametes of the hybrid independently of the combinations that existed in the gametes of the original parents. (Linkage is obviously an exception to this rule but it was not discovered until much later).

Mitosis. The ordinary process of nuclear division by which two daughter nuclei are produced each with the same number of chromosomes as the parental nucleus.

Monohybrid cross. See Cross.

Monohybrid ratio. The theoretical 3:1 ratio of the numbers of the two different phenotypes obtained from a monohybrid cross showing dominance. This ratio is modified to 1:2:1 when dominance is not shown.

Monokaryon. A fungal mycelium consisting of cells each containing only one nucleus.

Mutation. An abrupt inheritable change expressed in a phenotype which may be due to a molecular change in a gene or to a loss, duplication or rearrangement of sections of chromosomes or even whole chromosomes.

Somatic mutation. A mutation that occurs in a somatic cell and which is in consequence not normally inheritable.

Mutant. An individual or type of individual with a phenotype different from the normal or wild type individual owing to the occurrence of a mutation.

Nucleic acids. Complex organic compounds found in the nuclei and cytoplasm of plant and animal cells and composed essentially of four nitrogenous bases, a sugar and phosphoric acid.

Deoxyribonucleic acid (DNA). The nucleic acid found almost exclusively in the cell nucleus and closely identified with the chromosomes.

Ribonucleic acid (RNA). The nucleic acid found in the ribosomes of the cytoplasm and which takes on several forms according to the function it performs in the synthesis of proteins. These forms are known as acceptor, messenger and transfer RNA respectively.

Parents. The two individuals or kinds of individuals used at the beginning of a breeding experiment.

Parental generation. All the individuals used as parents in the first generation of a breeding experiment. Denoted symbolically as the P or F_0 generation.

Parental types. The genotypes and phenotypes of the two kinds of parents used in a breeding experiment.

Phenotype (-typically). The appearance of an organism with respect to a particular character or combination of characters resulting from its genotype and its interaction with its internal and external environment.

Pleiotropy (-tropic). Multiple phenotypic effects of a single gene.

Polyploid. Having three or more complete sets of homologous chromosomes. See also haploid and diploid and chromosome number.

Progeny. The offspring in the first and succeeding generations of a breeding experiment.

Prototroph (ic). A term used in fungal and bacterial genetics for the normal or wild type organisms that can grow on minimal nutritive medium.

Pure line (pure bred). An inbred series of individuals which when mated together breed true with respect to the particular characters under consideration.

Recessive. The term used by Mendel to describe a character from a contrasting pair of characters used in a breeding experiment which was not expressed by the individuals in the F_1 generation but which reappeared in one quarter of the individuals in the F_2 generation. See also dominant.

Reciprocal cross. See cross.

Recombination (Recombinant). The processes by which new combinations of parental characters arise in their offspring. Recombination can occur in several ways but the term is of particular use in the study of linkage where recombination follows from crossing-over. Thus in the cross denoted by

$\frac{AB}{AB} \times \frac{ab}{ab}$, the test cross would be $\frac{AB}{ab} \times \frac{ab}{ab}$ and the recombinants $\frac{Ab}{ab}$ and $\frac{aB}{ab}$.

Repulsion. The condition in a heterozygote carrying two linked characters in which one dominant character and one recessive character are inherited from one parent and the corresponding recessive and dominant characters from the other. This condition is represented symbolically as $\frac{Ab}{aB}$. See also coupling.

Scoring. Counting the numbers of individuals of the different kinds of phenotypes in the succeeding generations of a breeding experiment.

Segregation. The separation of the allelic differences one from another. Segregation may arise as the result of the random orientation of the chromosomes at metaphase of the first meiotic division or, as the result of crossing over, at the second meiotic division.

Selection. The process of picking out those members of a population which will be the parents of the next and subsequent generations which leads ultimately to changes in the composition of the population and in due course to evolution.

Natural selection. Selection as it occurs under natural conditions uninfluenced by the deliberate action of man.

Artificial or human selection. Selection by the conscious action of man.

Self colour. The term used to describe the condition in which the colour of an animal, or of a flower, is the same or very nearly the same all over the animal or flower.

Species. An important biological concept that is fairly easily understood in principle but very difficult to define precisely. A group or category of plants or animals which can usually interbreed and have sufficient in common to enable them to be classified together as above one level of similarity but below another.

'Split'. A term used, particularly in the genetics of cage birds, to indicate a heterozygote carrying a recessive character which is masked by the presence of the dominant. Thus, as an example taken from the genetics of budgerigars, light green/blue, read as light green split blue, means a heterozygote which is light green in phenotype but which has the genes for both light green and blue in its genotype.

Variation. The differences to be found within a species or a population.

Continuous variation. Quantitative differences, for example, height in man, such that individuals differ from each other along a continuum. The difference between one member and the next in order being small and not susceptible to analysis by breeding experiments.

Discontinuous variation. This is marked by distinct qualitative differences. For example, the blood groups in man, which can be analysed by breeding experiments.

Environmental variation. Differences caused solely by the action of environmental factors so that individuals of the same genotype appear different because they have been exposed to different environmental conditions. These are not inheritable.

Genetical variation. Inheritable differences resulting from the different genetic constitutions (genotypes) of the individuals concerned. It should be noted however that the expression of a genetic character always depends on interaction with an appropriate environment.

Wild type. A term used to describe the normal individual of a species, or of a population, with respect to the character or characters under consideration, as distinct from a mutant.

Zygote. A cell formed by the union of two gametes.

Notation for crosses

Crosses may be represented symbolically or by describing the phenotypes in full. The following crosses for the mouse may be taken as typical examples. By convention the female parent is given first.

Monohybrid crosses

1 P Parents	BB Homozygous black female	x crossed with	bb Homozygous chocolate (brown) male
G Gametes	B (only one of a pair of characters can be represented in a gamete)		b
F_1 First filial generation		Bb All heterozygous black (black B is dominant)	

2 P_2	Bb Female Both heterozygous black. Individuals from the F_1 of the previous cross.	x	Bb Male
G	B and b		B and b

F_2	BB :	Bb :	Bb :	bb
	Black	Black	Black	Chocolate
	Homozygous	Heterozygous		Homozygous
	3 Black		: 1	Chocolate

These ratios are only obtained if the different kinds of gametes are produced in the same proportions from each sex and they have the same chance of fusing at fertilisation.

The fact that these ratios are obtained, with few exceptions, throughout the plant and animal kingdoms provides strong evidence that these conditions do operate.

Reciprocal cross of 1 would be:

3 P	bb Homozygous chocolate female	x	BB Homozygous black male
G	b		B
F_1		Bb All heterozygous black, as in 1	

Back cross from 1 would be the following or its reciprocal:

4	P_2	Bb Heterozygous black female (F_1 individual)	x	bb Homozygous chocolate male (the male parent from 1 or a male with a similar genotype)
	G	B and b		b
	F_1	Bb Heterozygous black	:	bb Homozygous chocolate
			1 : 1 ratio	

Dihybrid crosses

5	P	BB,DD Homozygous black female	x	bb,dd Homozygous fawn (dilute chocolate) male
	G	BD		bd
	F_1		Bb, Dd All heterozygous black (black B and full colour expression D are dominant)	
6	P_2	Bb,Dd Female	x	Bb,Dd Male

Both heterozygous black. Individuals from the F_1 of the previous cross.

G BD : Bd : bD : bd

Gametes from both parents will be as above, theoretically produced in equal proportions.

F_2		Gametes from male BD	Bd	bD	bd
Gametes	BD	BB,DD	BB,Dd	Bb,DD	Bb,Dd
from	Bd	BB, Dd	BB,dd	Bb,Dd	Bb,dd
female	bD	Bb,DD	Bb,Dd	bb,DD	bb,Dd
	bd	Bb, Dd	Bb,dd	bb,Dd	bb,dd

Genotype		Ratio	Total	Phenotype
BB,DD	:	1	9	Black
BB,Dd	:	2		
Bb,DD	:	2		
Bb,Dd	:	4		
BB,dd	:	1	3	Blue-grey
Bb,dd	:	2		
bb,DD	:	1	3	Chocolate
bb,Dd	:	2		
bb,dd	:	1	1	Fawn

This is the dihybrid ratio of 9:3:3:1

Reciprocal cross of 5 would be:

7 P	bb,dd Homozygous fawn (dilute chocolate) female	x	BB,DD Homozygous black male
G	bd		BD
F_1		Bb,Dd All heterozygous black	

Back cross from 1 would be the following or its reciprocal:

8 P	Bb,Dd Heterozygous black female (F_1 individual)	x	bb,dd Homozygous fawn (dilute chocolate) male (the male parent from 1 or a male with a similar genotype)

G	From female BD	:	Bd	:	bD	:	bd
	From male bd						
Backcross progeny	BD,bd Black	:	Bd,bd Blue-grey	:	bD,bd Chocolate	:	bd,bd Fawn

All the F_2 types from 2 in a 1:1:1:1 ratio

The statistical treatment of results

It will probably be during an elementary course in genetics that pupils will first encounter the problem of judging to what extent observed results which differ from expected results may be due to chance or point to some significant difference that requires further investigation [2,22]. In the basic course an empirical approach of tossing coins, throwing dice or cutting cards to represent the results obtained from breeding experiments will carry sufficient conviction and introduce the pupil to the fundamental ideas but in the extension course the use of the χ^2 (chi-squared) test, at least as a tool, will be essential.

The χ^2 test was devised by Karl Pearson in 1889. It has become a powerful statistical device for testing whether or not results obtained from an experiment, which involves sampling, agree with a given hypothesis. To answer this question two pieces of data are needed; a measure of the deviation of the sample from the hypothetical population and a means of judging whether or not this measure is an amount commonly experienced in sampling, or, on the contrary, is so great as to throw doubt on the hypothesis. χ^2 is a measure of the deviation and is an index of dispersion. Its probable occurrence in sampling can be obtained from χ^2 tables drawn up originally by Pearson.

χ^2 = the sum of all values of $\frac{(O-E)^2}{E}$ where O is the observed or sample frequency and E is the expected or theoretical frequency. Squaring eliminates the sign of the deviation and is a common device in statistics. The occurrence of the expected numbers in the denominator introduces the size of the sample into the index.

Probability (p) is expressed quantitatively on a scale from O, which represents an impossible event, to 1, which represents a certain event. (Sometimes these are multiplied by a hundred and expressed as a percentage). A probability of 0.5 (½) means that on average the event will occur once in every two trials, like calling heads when tossing coins and a probability of 0.25 (¼) means that on average the event will occur once in every four trials, like drawing a spade when drawing a card from a pack. The probability of throwing a six with a single dice is 0.17 (1/6).

The χ^2 test gives the probability of obtaining a deviation from the expected as great or greater than the observed deviation by chance alone. Whether this deviation can be considered as significant or not is a matter of judgment when all the factors have been taken into consideration. It is customary to regard a probability of 0.05 (1/20) as significant and one of 0.01 (1/100) as highly significant but it must be emphasised that these values are only indicative and each case should be considered in detail on its merits.

The following examples should make the use of the χ^2 test clear:

1 Monohybrid cross with mice: black (dominant) x brown (recessive). Monohybrid ratio in offspring of 3 black to 1 brown expected.

Pair	No. of litters	Observed		Expected	
		Black	Brown	Black	Brown
A	8	35	17	39.0	13.0
B	6	36	14	37.5	12.5
Combined	14	71	31	76.5	25.5

For Pair A alone $\chi^2 = \frac{(35-39)^2}{39} + \frac{(17-13)^2}{13} = 1.64$

Pair B alone $\chi^2 = \frac{(36-37.5)^2}{37.5} + \frac{(14-12.5)^2}{12.5} = 0.24$

Combined $\chi^2 = \frac{(71-76.5)^2}{76.5} + \frac{(31-25.5)^2}{25.5} = 1.60$

Pair A alone p is between 0.25 ($\chi^2 = 1.32$) and 0.10 (2.71) By interpolation p = 0.28
Pair B alone " 0.75 ($\chi^2 = 0.10$) 0.50 (0.45) p= 0.65
Combined " 0.25 ($\chi^2 = 1.32$) 0.10 (2.71) p = 0.22

The next step is to find the probabilities from a chi-squared table and this requires calculating the number of the 'degrees of freedom'. When a single set of numbers is being compared with known or expected proportions, this is one less than the number of terms in the expression for χ^2, which in this case is two, and hence the number of degrees of freedom is one.

A better method in a case like this is to treat the samples separately, rather than to combine the results. The two values for χ^2 are added and the table entered for two degrees of freedom, the number of samples included.

Thus $\chi^2 = 1.64 + 0.24 = 1.88$ for two degrees of freedom : $p = 0.41$

This shows that the two samples when treated separately support each other and give a stronger indication that the hypothesis is supported by the results.

2 Dihybrid cross with maize: coloured starchy (double dominant) x white sugary (double recessive). Dihybrid ratio in offspring of 9:3:3:1 expected as shown.

	Coloured Starchy		Coloured Sugary		White Starchy		White Sugary	
	Observed	Expected	Observed	Expected	Observed	Expected	Observed	Expected
No of grains	2862	2966	993	989	1055	989	365	330
Ratio	8.70	9	3.02	3	3.20	3	1.11	1

Note. The expected numbers are calculated from the theoretical ratios and the total grains counted. Thus the expected number of coloured starchy grains is obtained from $\frac{9}{16} \times 5275 = 2966$.

$$\chi^2 = \frac{(2862-2966)^2}{2966} + \frac{(993-989)^2}{989} + \frac{(1055-989)^2}{989} + \frac{(365-330)^2}{330}$$

$$= \frac{(-104)^2}{2966} + \frac{4^2}{989} + \frac{66^2}{989} + \frac{35^2}{330} = 3.65 + 0.02 + 4.40 + 3.71$$

$$= 11.78 \text{ for 3 degrees of freedom.}$$

From the tables p is less than 0.01 and hence the departure from the expected result is highly significant. A re-examination of the data shows that there was an excess of white grains or conversely a shortage of red grains so that an explanation, if one was to be found, must be looked for in that direction. This investigation was in fact carried out in a school using the method described on page 59. The method requires that all the 'white' plants are emasculated. It could have been that a 'white' plant was planted in a 'red' row inadvertently and thus escaped emasculation or, more likely, that a secondary male inflorescence developed after the school holidays began and produced some pollen before it was detected. The point is that good data should never be discarded; much can be learned from cases where unexpected results are obtained.

2 Vertebrate animals–mammals

Studies with mammals probably provide the best starting point for any genetics course, be it elementary or advanced. There are many other benefits besides those bearing on genetics and provided adequate records are kept these can provide material for discussion. Mammals kept in schools should therefore wherever possible be those varieties which can be used for genetics.

Recording
Each cage should carry a label, indicating the sex and characteristic of the parents and the nature of the mating, for example, the purpose of the investigator or the production of certain stock. Two pages in a record book for each investigation are useful, one for the mating records and the other for the litter records. A code indicating the nature of the experiment and the stage it has reached can easily be devised. A simple code system especially suitable for mice is given in reference 70.

Management
Essential details of the management techniques for small mammals are given in Table 13, page 90. Full details are given in the companion publication *Small Mammals*[36] which should be consulted for further information. All animals must be obtained from reliable sources, preferably from breeders accredited by the Medical Research Council Laboratory Animals Centre.

The numbers in *italics* after the heading for each organism refer to the points indicated on the 'outline scheme' Fig 1 and Table 1.

Mouse, Mus musculus (2n = 40) *1, 4, 12, 14.*

Genetically more is known about the mouse than any other mammal; it lends itself to breeding experiments, well over a hundred mutant genes having been identified and related to the twenty linkage groups (see Table 3). Mice tolerate handling and examination and the many coat colour mutants are very distinctive and easy to score soon after birth.[39, 41, 70, 71]

It is wise to begin breeding mice for genetics on a small, unambitious scale and expand as experience grows. If good records are kept there is no need to have all the desired investigations running at the same time so long as each student has full responsibility for one for long enough to appreciate the system of management, the various techniques and the method of recording.

Coat colour
Most of the genes suitable for elementary work affect coat colour and some interesting preliminary studies can be made into the nature and distribution of the colouration. The colours of the coats of mammals are due to pigment granules in the hairs, and there is a considerable similarity between those of different mammals. The granules are of two forms: eumelanin, when they may be black or brown, and phaeomelanin, when they may be yellow or reddish. The colour of the coat is determined by the relative frequency of the various granules, their density and their distribution in the hairs. Hairs can be plucked from different regions of the bodies of mice and other mammals of different coat colours and examined under the microscope. The coat colours were given names by the Mouse Fancy before the mechanism of inheritance was known.[49,60]

Suggested investigations (See Table 4)

1 *Monohybrid crosses*

1.1 *Monohybrid ratio (3:1) and backcross (1:1)*
black (BB) x chocolate (bb), reciprocal and backcross
black (BB,DD) x blue-grey (BB,dd), reciprocal and backcross
black (BB,PP) x blue-lilac (BB,pp), reciprocal and backcross
black (AA,BB) x tan-belly (a^ta^t,BB), reciprocal and backcross

Table 3 Mouse mutants suitable for school studies

Linkage Group		Symbol	Description of the mutant	Comments and Suggested Use
1	extreme chinchilla	c^e	pale coffee colour with black eyes	Two members of an allelomorphic series which includes the wild type gene C for full colour. Modified (1:2:1) monohybrid cross and backcross.
	albino	c	familiar pink-eye white mouse	
	pink-eye dilution	p	pink eyes and sandy coat	
2	Maltese dilution	d	blue-grey if B present dilute-chocolate with bb	Monohybrid and dihybrid crosses and backcrosses.
5	Yellow lethal	A^y	yellow in heterozygous A^yA condition	Homozygotes A^yA^y die as embryos. Four members of an allelomorphic series found in several other mammals suggesting homologies (see guinea pigs and rabbits). Monohybrid and dihybrid crosses and backcrosses.
	Agouti	A	wild mouse coat	
	black and tan	a^t	black with a tan belly	
	non-agouti	a	black self coloured unless diluted by a dilution gene	Sd gene seriously affects skeletal and urinogenital systems. Sd Sd homozygotes die. These three genes can be used for linkage investigations.
	Danforth's short tail	Sd	ordinary wild mouse with a short tail	
	Wellhaarig	we	wavy coat	
8	brown (= chocolate)	b	brown self coat in the absence of A	Brown eumelanin produced instead of black. Monohybrid and dihybrid crosses and in combination with a, Sd and we for linkage investigations.
13	leaden	ln	very close to d above	Linkage
	fuzzy	fz	hair thin and faintly wavy	
20	Tabby	Ta	TaX/–Y ♂ and TaX/TaX ♀ have greasy fur, bare patches behind ears and little hair on the tail and the so-called 'tabby' coat. TaX/+ X ♀ has a less pronounced tabby coat	Sex linkage and pleiotropic action of gene. The tabby gene is located on the nonhomologous part of the X chromosome.

1.2 *Monohybrid ratio with incomplete dominance (1:2:1)*
albino (aa,cc) x extreme chinchilla (aa,c^ec^e), reciprocal and backcross
This is a particularly useful investigation since the heterozygote c^ec is black-eyed white so that if the F_2 progeny are scored on coat colour alone or eye colour alone 3:1 ratios are obtained, emphasizing the point that the 3:1 ratio is a special case of 1:2:1 ratio rather than the reverse.

1.3 *Monohybrid ratio with lethal homozygote (2:1)*
yellow (A^yA) x yellow (A^yA) gives 0:2:1

Table 4 The relationships of the self colours of the mouse as determined by the genes a (non-agouti), b (brown), c (absence of colour or albino), d (Maltese dilution) and p (pink-eyed dilution).

Series	Type	Genetic Formula				
Agouti	Agouti or wild type	AA	BB	CC	DD	PP
black	black	aa	BB	CC	DD	PP
	blue-grey	aa	BB	CC	dd	PP
	blue-lilac (dove)	aa	BB	CC	DD	pp
chocolate	chocolate (brown)	aa	bb	CC	DD	PP
	dilute chocolate (fawn)	aa	bb	CC	dd	PP
	champagne	aa	bb	CC	DD	pp
albino	albino (any mouse without the C factor but here it is given in its pure form and hence valuable for genetical testing)	aa	bb	cc	dd	pp

2 *Dihybrid crosses without linkage*

2.1 *Dihybrid ratio (9:3:3:1) and backcross (1:1:1:1)*
black (BB,DD) x fawn (bb,dd), reciprocal and backcross
black (BB,PP) x champagne (bb,pp), reciprocal and backcross
chocolate (bb,DD) x blue-grey (BB,dd), reciprocal and backcross
The following two investigations introduce the idea of 'coupling' and 'repulsion' without linkage:
pink-eye (pp,SeSe) x short-ear (PP,sese), reciprocal and backcross
pink-eye, short ear (pp,sese) x wild type (PP,SeSe), reciprocal and backcross
This is a useful investigation affecting as it does two quite distinct characters, since the F_2 progeny can be scored for eye colour alone or ear shape alone to give the 3:1 monhybrid ratio, thus emphasising the independent segregation of the two characters.

2.2 *Dihybrid ratio with epistasis (9:3:4)*
black (BB,CC) x albino (bb,cc), reciprocal and backcross

3 *Linkage*

3.1 *Sex linkage.* In the mouse, as in almost all mammals, the male is the heterogamous sex (XY). The tabby gene Ta is in the non-homologous part of the X chromosome so that the heterozygous male TaX/—Y has the greasy appearance of the TaX/TaX female. (See Table 3)

	P	tabby female TaX/TaX x agouti male +X/—Y
gives	F_1	tabby females TaX/+X and tabby males TaX/—Y in equal proportions
	P	agouti female +X/+X x tabby male TaX/—Y
gives	F_1	tabby females TaX/+X and agouti males +X/—Y again in equal proportions

tabby indicates the heterozygous female condition in which a faint striping appears.

3.2 *Autosomal linkage.* The general characteristics of linkage with some cross-over can be shown: leaden (lnln,++) x fuzzy (++, fzfz), both in linkage group X111 and, leaden fuzzy (lnln, fzfz) x wild type (++,++), the heterozygotes obtained in each case being backcrossed to leaden fuzzy.

Other possible investigations include:

4 *Chromosome mapping and interference*
This can be shown with Danforth's Short tail (Sd) Wellhaarig (we), non-agouti (a) and brown (b). For further details see references 70 & 71.

5 *Modifying genes and polygenic inheritance*
This can be shown with polydactylous hind feet.

6 *Quantitative inheritance*
This can be studied in relation to body weight. For further details see references 70 and 71.

Rabbit, Oryctolagus cuniculus (2n = 44) *4, 16*

Rabbits are not very suitable for genetical studies in schools and their use for genetical investigations alone is difficult to justify. Where they are kept for other purposes however, the opportunity can be taken to investigate various genetical problems.[35, 59]

Coat colour

The coat colour mutants are very similar to those of the mouse[60]:

A series: A agouti; A^w light bellied agouti; a^t black and tan, a non-agouti

B series: B black; b brown

C series: C full colour; c^{ch} chinchilla; c^h Himalayan; c albino

In the rabbit, the B and C series are linked, whereas in the mouse they are on different chromosomes.

D series: A recessive mutant (d), which dilutes full colour, is known but it is doubtful whether it is homologous to the D series in the mouse.

White-spotting: Dutch spotting, (du), in its typical form consists of a wide belt of white encircling the front half of the body including the fore-legs. English spotting, due to a dominant (En), is very variable in extent and may range from almost pure black through every grade of intermediate to almost pure white.
These two genes are closely linked with a cross-over frequency of only 0.17%.

Genes affecting hair structure

There are a number of these and the following selection provides a suitable range for school studies when combined with the colour genes.

Angora (l) greatly increases the length of the hair and also its composition. Rex (r_1), (r_2) and (r_3) are three quite distinct genes which shorten the hair to give a fine very smooth coat. r_1 and r_2 are linked.

Rabbit breeds

Beveren: dense coats due to the action of polygenes. There are several colour varieties, the best known being the Blue Beveren aa,dd.

Chinchilla: $c^{ch}c^{ch}$.

Dutch: dudu.

English: For exhibition these are usually heterozygous for En so that the amount of spotting is moderate; ground colour black aa,BB, Enen or brown aa,bb, Enen are both acceptable.

Flemish: basically dark, white-bellied agouti A^w/A^w

Havana: aa,bb with some dark modifiers.

Himalayan: the temperature sensitive c^h on a non-agouti background aa; they have dark extremities.

Sable: may be cc^{ch},$c^{ch}c^{ch}$ or c^hc^{ch}. They are also, to some extent, temperature sensitive.

Suggested investigations

1 *Monohybrid cross*
Monohybrid ratio (3:1).
black Dutch (BB) x brown Dutch (bb), reciprocal and backcross.

2 *Dihybrid cross*
Dihybrid ratio (9:3:3:1).
black self (BB,DuDu) x brown Dutch (bb,dudu), reciprocal and backcross.

3 *Environmental (temperature) sensitivity of the Himalayan gene*
Pure-bred Himalayan rabbits are pure white when born, their extremities (nose, ears, tips of feet and tail) only turn dark or black after leaving the warm nest. Measurement of the temperature of these extremities shows that they are below that of the normal body temperature which is about 38.5°C; the temperature of the nose may be as low as 25°C. A number of investigations may be undertaken into this effect. Will the young develop dark extremities if kept at a temperature of 30°C. What would be the effect of bottle feeding the

young as soon as they can be trained to take food in this way and rearing them at room temperature? How can these suggestions be improved? A useful beginning to further work is to take the skin temperature of a rabbit at various points by means of a sensitive thermistor thermometer.

4 *Multiple allelomorphs*
P Himalayan (bb,c^hc^h) x brown (bb,CC), reciprocal and backcross.
P albino (bb,cc) x brown (bb,CC), reciprocal and backcross.
F_1, in both cases, are all brown and for the F_2, 3 brown to 1 Himalayan (or albino as the case may be).
P Himalayan (bb,c^hc^h) x albino (bb,cc), reciprocal and backcross.
F_1, are all Himalayan and for the F_2, 3 Himalayan to 1 albino. Himalayan and albino are behaving as two allelomorphs to C and the explanation must be that they either occupy the same locus or are so close together in the same chromosome that there is perfect linkage between them. A good exercise would be to get students to design confirmatory investigations.

Guinea pig or cavy, Cavia porcellus (2 n = 64) *4*

The genetics of the guinea pig has been intensively studied and the animals make attractive pets but the long period of gestation (9-10 weeks), long generation span (6 months) and small size of litters (1-5) militate against its use for establishing genetical principles. When the animals are being used primarily for other purposes however there are clear advantages in making crosses that illustrate genetical principles in a similar way to the use of the rabbit.[64]

Suggested investigations

1 *Monohybrid crosses*
Monohybrid ratio (3:1)
short-hair (LL) x long-hair (Angora) (ll), reciprocal and backcross
rough-hair (rosetted) (RR) x smooth hair (wild type) (++ or rr), reciprocal and backcross
pigmented (actual colour unimportant) (CC) x Himalayan (c^hc^h); reciprocal and backcross.

2 *Environmental (temperature) sensitivity of the Himalayan gene*
For information, see the section on the rabbit.

If stocks of known genotypes are unobtainable or not used, then investigations to determine the genotype of unknown phenotypes can be carried out.

Syrian or golden hamster, Mesocricetus auratus (2n = 44) *4*

Mutants of this species so far identified are very similar to those found in the mouse, guinea-pig and rat.[55,60]

Suggested investigations

1 *Monohybrid cross*
Monohybrid ratio (3:1)
Golden (wild-type) A^wA^w, EE) x cream (A^wA^w,ee), reciprocal and backcross
The A^w gene is believed to be homologous with others in the A agouti series.
Golden (Wild-type) (A^wA^w, SS) x piebald (A^wA^w, ss), reciprocal and backcross.
When the chromosome number, 2n = 44, of the Syrian hamster is compared with that of the European hamster *Cricetulus cricetus*, 2n = 22, there is a strong suggestion that the Syrian hamster is a tetrapolid. Polyploidy is extremely rare in animals.

Laboratory rat, Rattus norvegicus (2n = 42), *4*

Suggested investigations

1 *Monohybrid cross*
Monohybrid ratio (3:1)
Hooded (Pigmented in head and shoulder region and with a mid-dorsal stripe) (CC hh) x albino (cc — —), reciprocal and backcross
Other colour genes suppressed by the absence of the colour factor in the albino may be revealed.

Mongolian gerbil, Meriones unguiculatus (2n = 44) *4*

Suggested investigations

1 *Monohybrid cross*
Monohybrid ratio (3:1)
Normal colour (Wild-type) (CC) x albino (cc), reciprocal and backcross
Albino gerbils have recently become available commercially.

Man, Homo sapiens (2n = 46) *1*

Boys and girls of all ages are naturally interested in how what they learn about inheritance from plants and animals applies to them. The course in genetics interwoven into the general biology programme, might in fact begin with a consideration of human variation and its resolution into those aspects which are continuous and those that are discontinuous; those which are predominantly inheritable and those that are largely environmental in origin (see Table 5). Similarly it might end with some assessment of the present situation and how a knowledge of genetics can help towards the achievement of a fitter, fairer and happier world.

Suggested investigations

1 *Characters showing continuous variation*
The following list gives a few easily measured characters which could prompt some simple attempts at standardisation:[5,12,18,21] height, weight, length of foot, length of forearm (elbow to finger tip), length of index finger, span, reach, pulse rate, cephalic index, acuity of vision and hearing, adaptation to low intensity light. Comparative results from sets of identical twins, fraternal twins, siblings at same age, relatives and so on can be accumulated and added to.

2 *Characters showing discontinuous variation*
Table 5 gives some characters known to be inherited in a fairly simple manner. For further information, see references 20, 21, 78, 81 and 82. To illustrate inheritance in man, family trees showing, for example, the inheritance of eye colour can be compiled by pupils for their own families. Human family histories often show striking inherited factors, for example, the transmission of haemophilia in the descendants of Queen Victoria.[21]

Notes on genetical characters in home and farm animals

Although these animals are not suitable for genetical studies in schools observations can often be made on them and the facts obtained related to information gained from other sources. One of the points that can emerge, and it is a very important one, is that there are striking patterns of genes, especially those controlling coat colour,[60] in mammals and that in general the closer the animals are taxonomically the closer the patterns seem to agree. This throws some light on the course of evolution and serves as a check in classification.

Cat, Felis catus (2n = 38)
A series represented by A agouti and a non-agouti
B series represented by B black and b brown
C series represented by C full colour and three dilutions, one of which is Siamese (This may be compared with Himalayan in rabbit)
D series represented by D intense and d dilute
The wild cat is tabby TaTa but this is not homologous with Ta in the mouse.
Short hair (LL) is dominant to long Persian hair (ll). White coat is not albino; the white cat has blue eyes and the trait is due to a dominant gene W.
Orange (sex linked) O, the colour of the marmalade cat, is interesting especially when considered in association with agouti (A,Ta) and non-agouti (a,Ta) since differences in the non yellow areas are produced.

In males OX/—Y gives orange or yellow coat, in heterozygote females OX/+X produces tortoiseshell (calico) coat and in homozygous females OX/OX produces, as expected, orange coat. The non-yellow areas are agouti in the tabby tortoiseshell and non-agouti in the ordinary tortoiseshell even when A is present.

There remains the question of how the occasional (rare) male tortoiseshell cat can be accounted for. These are usually sterile and it is now generally

Table 5 Human variation, characters known to be inherited.

Character	Phenotypes and Genotypes	Remarks
Type of hair 1	Curly c^1c^1 straight c^2c^2 wavy c^1c^2	
Type of hair 2	Kinky S– non-Kinky ss	
Colour of hair 1	Dark D– light dd	This enables a rough classification to be made but is obviously over-simplified. Many genes are undoubtedly involved.
Colour of hair 2 (red pigment)	Absent R– present rr	
Colour of eyes	Pigmented E– non pigmented ee	There are other genes acting as modifying factors
Ear lobes	Free L– attached ll	
Ability to roll tongue	Roller A– non-roller aa	
Ability to taste phenyl thiocarbamide (PTC), or phenyl thiourea (PTU)	Taster T– non taster tt	Only suitably prepared paper (1cm x 2.5cm strips soaked in 0.13% PTC aqueous solution) must be used; on no account must solutions be used. See references 9 & 10.
Ability to smell different strains of freesias		
Mid-digital hair	Present M– absent mm	
Double jointed thumb	Present J– absent jj	
Presence of long palmar muscle	Present ww absent W–	
Crooked little finger	Bent P– straight pp	
Blood grouping	O, A, B, AB Rh^+, Rh^-	There is a risk to health in blood sampling. Only sterile lancets should be used on areas of skin swabbed with surgical spirit. It is advised that reference 9 and 10 be consulted before an attempt at blood sampling. 'Eldon' cards specially made for blood grouping could be used.
Haemophilia Red/green colour blindness	totally sex-linked recessive genes	

believed that they are XXY in genetical constitution resulting from non-disjunction in the formation of the male gametes, which gives XX and O gametes, and thus XXY and XO after fertilisation.

One further interesting point about the genetics of the cat is the inheritance of the short tail of the Manx cat. Manx cats are heterozygous T^aT for the semi-dominant tailless condition. TT x T^aT^a segregates 1 TT (tailed): 2 T^aT (Manx): 1 T^aT^a (tailless). Breeders usually cross Manx males with tailless females (the backcross) to get 1 (Manx): 1 (tailless).

Cattle, Bos taurus (2n = 60)

The evidence for the colour series given in the case of other mammals is not conclusive in cattle and details are not given here. Some coat colour inheritance however is well established and figures in most elementary books of genetics especially the earlier ones. Examples can often be found on farm visits.

red (RR), (e.g. Red Poll or Lincoln Red) x white (rr), (e.g. White Shorthorn) gives roan as 1:2:1 heterozygote.

black (WW), (e.g. Aberdeen Angus) x white (ww), (White Shorthorn) gives blue roan as 1:2:1 heterozygote.

Some white spotting genes of interest are:

Hereford pattern (Bl^h) which gives the white face often found in dual purpose cattle where a Hereford bull is used. The white face is regarded as an indelible stamp of Hereford in the blood.

Belted (Bt^1) found in Belted Galloway gives a white belt running round the body just behind the shoulder.

Recessive white spotting(s) found in Friesians (generally with black) and Shorthorns (with red) and Jerseys (with brown).

Black in Friesians is dominant to red. Formerly 'Red' Friesian calves were slaughtered but there is now a Red Friesian Society!

Hornless (polled) is dominant to horned. Now that cattle are often kept in closed yards and milking is done in streamlined milking parlours, polled cattle are more in favour and the hornless gene is being introduced into many breeds even including Ayrshires for which previously the set of the horns was an important show point.

The Dexter, a breed which has now almost disappeared, was interesting because it was a heterozygote for a lethal gene which in the homozygous condition gave what was known as a bull-dog calf, a very deformed creature born dead. Dexters had short legs and were small. They were maintained by crossing with the Kerry, a small black horned breed adapted to the wet climate of south west Ireland where it originated:

++ (Kerry) x +de (Dexter) gave 1 ++ (Kerry): 1 +de (Dexter)

Sheep, Ovis aries (2n = 54)

As in cattle the evidence on the colour series described in the mouse and commented on in the sections dealing with other mammals is limited and inconclusive.

Brown sheep, where the brown is recessive to black, are found in some primitive breeds such as the Soay.

Dominant black *and* brown occur in some breeds of Asian origin and it is thought that several genes may be involved. On the other hand there is evidence for two genes for which black is recessive to white, so the black/brown/white relationships in sheep are obviously very complicated. A dominant white gene distinct from the other genes referred to above is responsible for the almost universal white colouration of sheep of European origin.

When the question of colour in sheep is raised it is often interesting to refer to the story of Laban and Jacob and the division of their flocks of sheep and goats as told in the Old Testament, Genesis chapter 30. The misconceptions but also the element of genetic knowledge based on experience that becomes apparent can form the basis of useful discussion.

The inheritance of the horned condition in some breeds of sheep follows a rather unusual pattern known as *sex-limitation,* a better term might be sex-influenced. Thus in the Dorset breed both sexes have horns (HH) whereas in the Suffolk and Southdown breeds both sexes are hornless (hh). When the two breeds are crossed all the males in the F_1 are horned and all the females are hornless. When the F_1 hybrid sheep are crossed amongst themselves however, males occur in the ratio of 3 horned: 1 hornless and the females in the ratio of 1 horned: 3 hornless. This can be explained by postulating that the horned condition is dominant in the males but recessive in the females. A similar condition occurs in the trait baldness in man.

Pig, Sus scrofa (2n = 38)
Here again the evidence on the coat colour series is scanty. The most interesting genes for colour are:

(E^d): black dominant (Berkshire), ($e^J e^J$) red with black (Gloucester Old Spot and possibly Tamworth) and (ee) red (Duroc-Jersey and possibly Tamworth).
I: inhibitor of colour (Landrace and Yorkshires or Large Whites),
(I^d) dilution, i: intense.

white saddle (SS) (Essex and Wessex) is dominant to self colour (ss) (black Hampshires or red Duroc-Jersey).

Horse, Equus caballus (2n = 64).
The A series represented by (A) agouti or bay, a^t tan points or brown, (a) non-agouti.
B series represented by B black, b chestnut or brown
E series represented by E^d dominant extension of black, E normal black, e restriction of black
(W) dominant white, though this does not give an albino since the eyes are coloured.
(D) dominant dilution, not to be confused with the recessive dilution of the D series.
Piebald—dominant gene gives extensive white spotting without definite pattern.
The following probable genotypes may make the action and interaction of these genes clearer.
Bay: black agouti (A–, B–) Dark (mahogany) bay: (A–, B–, E–.) Red (blood) bay: (A–, B–, ee).
Intense black: (E^d–, B–.) Black: (E–, B–, aa.) Recessive black: mane and tail darker than the body, (aa, B–, ee).
Chestnut: (A–bb, E–) or (aa,bb, E–.) Light chestnut: (A–, bb, Dd, E–) or (aa,bb,Dd, E–).
Grey: Many combinations of genes can give greys due to G a mominant grey factor.
Palomino: (A–,bb, E–Dd) or (A–,bb,ee,Dd).
Piebald: any black horse with the piebald gene.
Skewbald: a non-black horse with the piebald gene.

In these examples of genetical characteristics of familiar mammals only those which are the expression of a single or, at most, two or three genes affecting discontinuous or qualitative characters have been given. More important characteristics from the point of view of commercial production like rate of maturing, milk yield, fleece quality, fat to lean proportion, are generally polygenic and continuous or quantitative and greatly influenced by environmental factors. This is what makes animal breeding for production difficult.

3 Other vertebrate animals– fish and birds

Fish (pisces)

Guppy, Poecilia reticulata (= Lebistes reticulatus) 2n = 46) *4, 9*

The Guppy is a member of the family *Poecilidae*, the viviparous toothed carps. The species is native to Central and South America and the West Indies where the fish live in pools, streams, drainage ditches and irrigation canals; their natural adaptability accounts for the ease with which they can be kept and makes them particularly useful for schools that have no special facilities.[40,45,76]

Management

See page 92

Suggested investigations
There are a number of sex linked colour genes and guppies obtained from pet shops usually possess an assortment of these so that stocks of known genotypes should be obtained from biological suppliers.

1 *Monhybrid (and dihybrid) crosses*
Two unlinked autosomal colour factors, for gold and blond respectively, are potentially suitable for monohybrid and in combination for dihybrid crosses. Both characters depend upon the size and distribution of the melanophores in the scales and the mutant genes, designated g and b, are recessive to their wild type alleles G and B. The gold homozygote has fewer but larger melanophores distributed largely round the perimeter of the scales so that the yellow body of the fish shows through giving it a dull gold appearance. The blond homozygote has much reduced melanophores, both in size and number, and the fish have a pale yellow colour. This genotype is also hard to obtain since the fish are more difficult to breed than the gold owing to their lower viability. For this reason most schools will probably limit their studies with these two genotypes to the monohybrid cross:
gold (gg) x wild type (GG), reciprocal and backcross.
For schools with considerable experience of rearing guppies, the dihybrid cross should be attempted if stocks of the blond homozygote can be obtained since differences in the viability of the various genotypes lead to departures from the expected 9:3:3:1 ratios and thus to the recognition that the Mendelian ratios are dependent upon certain assumptions that are not usually specified:
gold (gg,++) x blond (++,bb), reciprocal and backcross

2 *Sex linkage*
In guppies the female is the homogametous sex of constitution XX and the male the heterogametous with the constitution XY. Colour genes are carried on both the X and the Y chromosomes, those on the Y chromosome being of course only expressed in the male.[77]

2.1 *Y linked genes*
Three are useful and show father to son inheritance; they cannot be transmitted through the female. They are:
Maculatus (Ma) Black spot on dorsal fin, red patch on body below the fin
Armatus (Ar) Swordlike extension of lower border of the caudal fin, black and red spots on the sides of the body
Pauper (Pa) Red patch and iridescence on posterior part of the body

e.g. Wild type female (XX) x Maculatus male (XYMa)

F_1 Normal wild type female (XX) and Maculatus male (XYMa) 1:1 ratio. No Maculatus females are produced.

2.2 *X linked genes*
The most useful of these is:
Half-black (bl) Posterior end of body a dull

black. This gene is, of course, expressed in both males and females.
e.g. Wild type female (XX) x Half-black pauper male (XBlYPa)
F_1 Half-black female (XXBl) and pauper male (XYPa) 1:1 ratio.
If these offspring are then allowed to interbreed:
F_2 Wild type female (XX), Half-black female (XXBl), Half-black Pauper male (XBlYPa), Pauper male (XYPa) 1:1:1:1 ratio.

Birds (aves) (See Table 14, page 94)
Two types of birds useful in genetical studies are given as examples in this chapter, namely cage birds, illustrated by budgerigars and canaries, and farm stock, illustrated by domestic poultry. Other cage birds could be used, in particular the zebra finch, see reference 58. In all birds the female is the heterogametic sex.

Management
See Table 14, page 94

Budgerigar, Melopsittacus undulatus (2n = 58) *1, 4, 12*

The wild bird, common in Australia from the tropical north almost as far south as the southern coast, is a small green parakeet with many local names. The familiar name by which we know it has been derived from an aboriginal word Betcheroygah which is said to mean 'good food'.

The first living budgerigars to be brought to Britain arrived in 1840 and although some sub-species with paler green colouration have been recognised, the vast range of coloured forms have been derived from mutations, selection and hybridisation in captivity, largely during the last fifty years although some of the major mutations were obtained earlier.

Some knowledge of the life of these birds in the wild helps when aiming to provide the best conditions for them in captivity. They are a migratory species, very gregarious and often to be seen in flocks of many thousands. They nest in very close proximity to each other in holes in trunks of trees and especially in Mallee trees, which send up a cluster of a dozen or more bare trunks from a single rootstock. The holes have an entrance of 3-5 cm and are from 15-40 cm deep; no nesting material is used. Adults and young feed on small seeds. The eggs are laid in clutches of 4-8 and the incubation period is 18 days. Several families may be reared in a year; the cock feeds the hen when she is incubating and the young when they hatch.

Budgerigars have become favourite pets and many children have a surprising knowledge about them. Suitable foods and equipment can be obtained locally from pet shops. There is a considerable literature so that information and advice are readily available. In addition, the genetics of budgerigars has been intensively studied which enables a suitable programme to be planned within the resources available (see Table 6).[66,73] However, it should be noted that there is a risk of a transmissible virus infection called psittacosis from budgerigars, especially if they are, or have been in contact with, recently imported stock.[24] The risk from good quality home-bred birds is very low if they are maintained in good condition and prevented from coming into contact with wild birds, especially doves and pigeons.

Table 6 Budgerigar mutants suitable for school studies

Type	Symbol	Mutant type	Remarks
blue	b	sky blue	monohybrid, dihybrid with autosomal linkage
dark	D	olive green	partial dominant giving 1:2:1 light green: dark green: olive green with BB and sky blue: cobalt: mauve with bb.
ino	Xi	pure yellow	sex linked giving lut*ino* (yellow) with BB and albino (white) with bb.

Suggested investigations

Two genetical uses for these birds are suggested, a study of colour variation and the inheritance of colour in simple cases. In deciding upon this rather limited programme several factors have been taken into consideration, the relatively small number of offspring that are produced (about 12 a year), the gregarious habits of the birds, the space required and the cost.

1 *Colour variation*

The only pigments present in the plumage of budgerigars are melanin, which may appear brown if the pigment granules are dispersed, and a non-granular yellow pigment in the non-cellular cortex of the feather barbs. The green and blue colours of the birds, which provide in combination the basis of the wide colour range, are due to physical effects of white light when it strikes the barbs of the feathers associated with the yellow pigment in the cortex of the barbs. There is thus a considerable variation in colour which can be investigated.

2 *Colour inheritance*

A large number of mutations affecting colour have been identified and fixed in budgerigars. The three given in Table 6 are considered to be the most suitable for elementary genetics as they affect the basic ground colours and not the less easily recognised patternings. The wild type is light green in colour.

2.1 *Monohybrid cross*

light green (BB) x sky blue (bb), reciprocal and backcross

2.2 *Monohybrid cross giving modified (1:2:1) monohybrid ratio since Dark (D) is incompletely dominant*

light green (BB,dd) x olive green (BB,DD) cross, which can be used to show autosomal linkage.

2.3 *Autosomal linkage*

The alleles for blue (B and b) and for dark (D and d) are on the same chromosome, or more correctly in the same linkage group, and hence they can be used to demonstrate autosomal linkage.
Budgerigar breeders recognise two types of cross involving these two pairs of alleles according to whether the dark green or the cobalt form is used as the heterozygote for backcrossing:
Thus in Type 1, dark green (BD,Bd) is backcrossed to sky blue (bd,bd) and gives theoretically 43% sky blue, 7% cobalt, 43% dark green/blue, 7% light green/blue.
In Type 2, cobalt (bD,bd) is backcrossed to light green (Bd,Bd) and gives theoretically 43% cobalt, 7% sky blue, 43% light green/blue, 7% dark green/blue.
In each case the percentage of recombinant types is 14 which is therefore the cross-over value between the two pairs of alleles. There is a slight complication in carrying out a conventional linkage investigation with these alleles since the dark mutant character is only partially dominant. The typical crosses would be mauve (bD,bD) x wild type light green (Bd,Bd) when the two mutant genes are in coupling, and olive (BD,BD) x sky blue (bd,bd) when the mutant genes are in repulsion. In each case the F_1 hybrid would be phenotypically dark green but with the genotypes (bD,Bd) and (BD,bd) respectively. These hybrids would be backcrossed to mauve (bD,bD) the double mutant.

2.4 *Sex linkage*

Criss-cross inheritance can be demonstrated with two crosses.
Light green ♀ (BB,XY) x yellow ♂ (BB,XiXi)
F_1 yellow ♀ (BB,XiY) and light green ♂ (BB,XiX) 1:1 ratio
Yellow ♀ (BB,XiY) x light green ♂ (BB,XX)
F_1 light green ♀ (BB,XY) and light green ♂ (BB,XiX)
The albinos with Xi (see Table 6) on the light blue (bb) background can of course be used in a similar way.

Additional information on colour genes in budgerigars.

Colour series of alleles. There is a dominant gene C for full colour with a series of alleles reminescent of that for the mouse (see page 30). Thus c^d gives dilute colour, c^w the form known as clearwings and c^g the form known as grey wings.

Sex-linked genes. In addition to the ino gene (see Table 6) there are the following: X_c cinnamon, X_l lacewing, X_o opaline and X_s slate, though they are not at the same locus and they affect patterning as well as colour.

Canary, Serinus canarius (2n = 80) *2, 4, 16*

Suggested investigations

Canaries have been popular cage birds for longer than budgerigars and they can be used in much the same way as budgerigars in the teaching of genetics, although the colours are not so clear cut. There are, however, two additional uses to which canaries can be put. Both concern the new, or red, factor.[51]

The wild bird is green with distributed dark pigmentation, and it comes from the Canary Islands. It has been bred as a pet from 1478 when the Spaniards conquered the islands and most of the colours were originally developed from mutations by selective breeding.

The red canary was bred in Germany at some time before 1928 by hybridisation with a South American bird, the hooded siskin *(Spinus caculatus)* and thus a new colour was introduced into the breeding of canaries and into the competitive classes at shows.

The second point about the red factor is that its expression depends upon a sufficient supply of carotene in the food. This enables a good demonstration of the interaction between the genotype and the environment to be set up. Two comparable samples of birds, preferably from the same clutch or clutches, are reared under the same conditions except that the food of one sample is without carotene while the other has added carotene. Special proprietary brands of canary food rich in carotene can be purchased, e.g. Carophyll Red, but it is more convincing to use natural foods like carrots or ripe apricots.

The domestic fowl, Gallus domesticus (2n = 78) *4, 12, 14, 20*

Where poultry are kept a strong case can be made for teaching some genetics through them. Many schools keep poultry as part of their environmental studies programmes and most of these introduce an element of genetics in connection with the economically important sexing of day-old chicks by the sex-linked genes affecting plumage colour or other characteristics.[57,67,68]

The domestic fowl is believed to have evolved from the red jungle fowl *(Gallus gallus)* which is found in the belt running from India to Indonesia. If this theory is true it points to the extreme plasticity of the original stock since such a wide variety of breeds have been derived. The latter fact leads some workers to suggest that hybridisation with other living or extinct species have also played a part.

The fowl has been domesticated for about 3500 years, originally for religious purposes, sport (as cockfighting) and as a time-keeper before its use for meat and eggs. The processes by which the fowl was transformed by man from a wild bird laying seasonally and incubating fewer than 50 eggs a year, to a farmyard animal laying an average 260 eggs a year, and in extreme cases as many as 350, are central to the practice of applied genetics.

Suggested investigations and demonstrations

For all these investigations bantams, miniatures of the normal breeds, can be used with considerable saving in costs and space but the commercial value of any surplus eggs will be less.

The first three suggestions are all demonstrations that played a part in the early development of genetics in the first decade of this century and the fourth is an attempt to plan a sound teaching demonstration to illustrate and explain an important commercial practice.

1 *Incomplete dominance and 1:2:1 ratio in the F_2*
Black Andalusian x White Andalusian, and its reciprocal
The F_1 gives the so-called Blue Andalusian which is really speckled white. When Blues are crossed they give Black: Blue: White in 1:2:1 ratio.

2 *Gene interaction in the formation of the comb*
rose-combed e.g. Wyandotte or Sebright x pea-combed e.g. Brahma, have walnut combs in the F_1 and these when intercrossed give a theoretical ratio of 9 walnut: 3 rose: 3 pea: 1 single. It is remarkable that with this double recessiveness, those breeds with single combs provide the vast majority of both breeds and birds.

3 *Unusual ratio of 13:3 in the F_2*
dominant white, e.g. Leghorn x recessive white, e.g. Wyandotte
White in poultry, as in some flower colours and in mammals, can be due to a colour inhibitor, when a colour factor is present or to the absence of a colour factor. The white of the White Wyandotte is an example of the first, where white is dominant, and that of the White Leghorn of the second, where white is recessive. When the two different whites are crossed the F_1 offspring are all white owing to the dominant inhibitor but when intercrossed for the F_2 whites and colours appear in the ratio of 13 white to 3 coloured.

P Leghorn (II,CC) x Wyandotte (ii,cc)
F_1 Ii,Cc intercrossed

F_2

Gametes	IC	iC	Ic	ic
IC	II,CC	Ii,CC	II,Cc	Ii,Cc
iC	Ii,CC	ii,CC	Ii,Cc	ii,Cc
Ic	II,Cc	Ii,Cc	II,cc	Ii,cc
ic	Ii,Cc	ii,Cc	Ii,cc	ii,cc

coloured

I dominant inhibitor.
C dominant colour factor.

4 *Sex linkage and criss-cross inheritance*
This and similar sex-linked characters such as bar plumage are used to sex chicks soon after hatching and so save the expense of rearing the unwanted cockerels. Silver (S) in breeds such as the Light Sussex is dominant to gold (s) in coloured breeds like the Rhode Island Reds. If the cross is carried out reciprocally the following results are obtained:

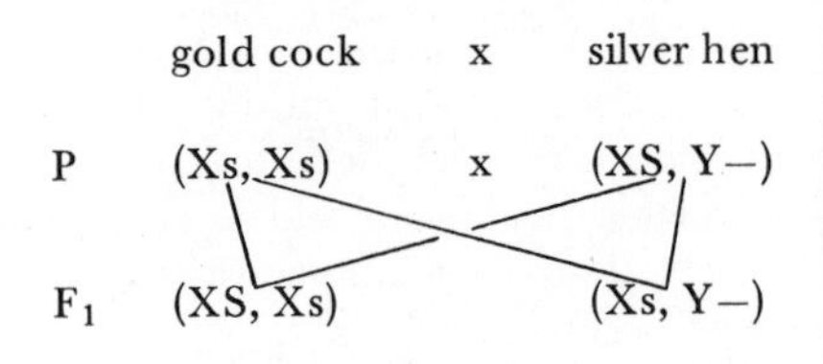

silver cockerels: gold pullets;
coloured chicks are all pullets
silver chicks are all cockerels

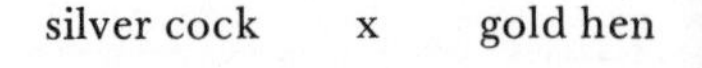

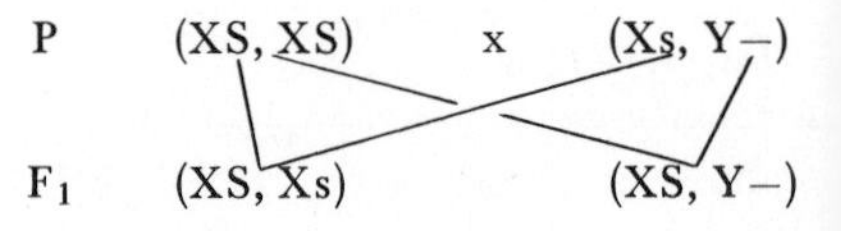

silver cockerels: silver pullets;
no distinction in plumage between cockerels and pullets.

Thus to be able to use sex linkage to sex chicks the cock must be golden and the hen silver.

Ideally, from a teaching point of view, two breeding pens of the two parental breeds are needed to show that they both breed true and two pens with the reciprocal crosses.

4 Invertebrate animals

Management

See Table 15, page 96.

Fruit or vinegar fly, Drosophila melanogaster (2n = 8)
1, 2, 4, 11, 12, 13

Of all the invertebrate animals used for genetics *Drosophila melanogaster* is without doubt the best known. This species was first used by an American, W.E. Castle, in 1906 and formed the basis of the breeding work carried out by T.H. Morgan and his team of brilliant investigators from 1909 onwards. Thousands of mutants have been discovered or induced and the literature on Drosophila genetics reaches astronomical proportions. The success of this tiny fly as an experimental organism for genetics lies in its ease of culture, short life cycle, small size, fecundity, the easily recognisable and hence scorable mutants and finally its small chromosome number.

Different species of *Drosophila* are common throughout the world in temperate and tropical regions. They are to be seen round ripe fruit in shops and markets or anywhere that fermented liquids are found.[32,61]

Table 7 Drosophila *mutants suitable for school studies*

Chromosome	Map Distance	Type	Symbol	Wild Type Characters	Mutant Character	Suggested Use
		Wild type	+	Grey body, red eye, long wings		most crosses
1 (sex)	0.0	yellow body	y	Grey body	yellow body	monohybrid, tight linkage
	1.5	white eye	w	Red eye	white eye	sex linkage, tight linkage, autosomal linkage
	36.1	miniature	m	Long wings	short wings*	autosomal linkage
	57.0	Bar eye	B	Normal eye	Narrow eye	temperature effect
2	13.0	dumpy wing	dp	Long wings	truncated wing*	monohybrid, dihybrid, populations
	67.0	vestigial wing	vg	Long wings	vestigial wing*	monohybrid, dihybrid
	104.5	brown eye	bw	Red eye	brown eye	monohybrid, gene interaction
3	44.0	scarlet eye	st	Red eye	scarlet eye	gene interaction
	50.0	curled wing	cu	Long wing	upturned wing*	autosomal linkage
	70.7	ebony body	e	Grey body	black body	monohybrid, dihybrid, autosomal linkage, populations

*See figure 2 for illustrations of the wing types.

Mutants suitable for school studies

The simplest characters to score are those affecting body colour, eye colour and wing shape. Details of a representative sample of these which enable the basic experiments to be carried out are given in Table 7. 14,30,33,37,44

Suggested investigations and some other uses

1 *Variation within a species*
The maintenance of as many mutant strains as possible together with wild type serves this purpose and the list given above, including as it does mutations affecting body colour, eye colour and wing structure (Fig. 2), would make a good beginning. Others affecting very minor characteristics such as bristle number or bristle distribution could be used to illustrate the wide range of structures affected and quantitative aspects.

2 *Interaction of environment and genetic constitution: the effect of temperature on the expression of the Bar eye gene*
Wild type flies have nearly a thousand facets in their compound eyes. This number is markedly reduced in a Bar eye individual, in which the eye appears as a narrow vertical strip. But the size of the strip, which is related to the number of facets and can therefore be made quantitative, varies between the sexes, being greater in the male, and with temperature, the higher the temperature the smaller the number of facets, that is the greater the expression of the gene. At a temperature of 15°C the average number of facets in the male is 270 and in the female 214; at 30°C 74 and 40 respectively.
This lends itself to profitable discussion on how the investigations should be set up, what hypotheses can be formulated to explain the results and how the hypotheses can be tested by experiment.

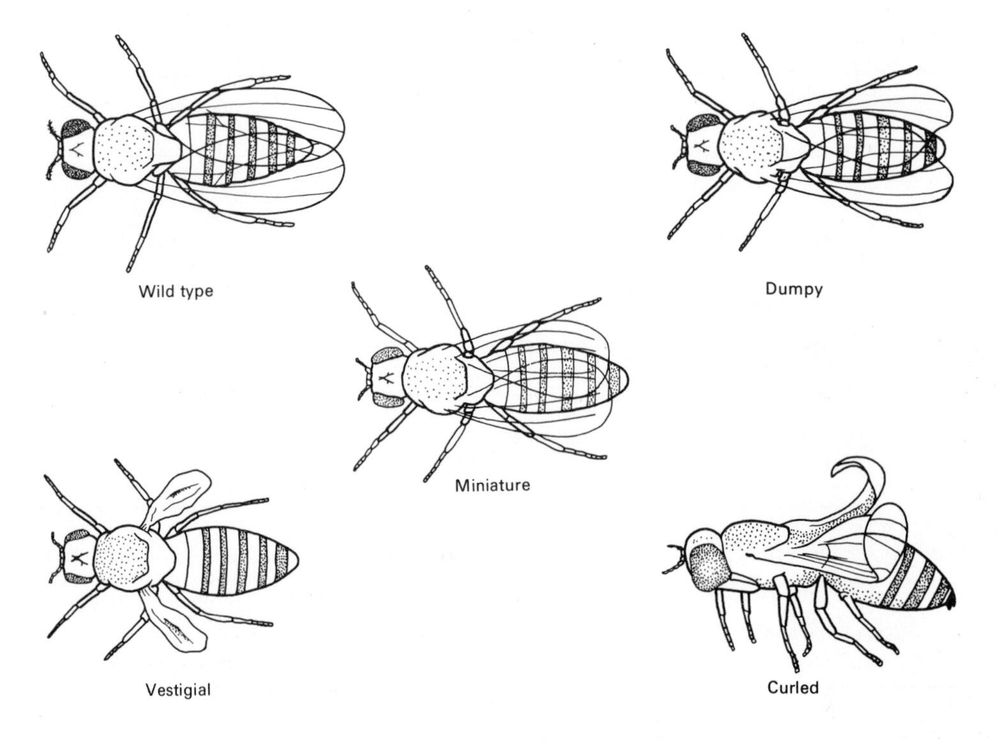

Figure 2 Wing types in Drosophila

3 *Monohybrid cross*

yellow body	(yy)	x	wild type (++) reciprocal and backcross
ebony body	(ee)	x	wild type (++) reciprocal and backcross
brown eye	(bw bw)	x	wild type (++) reciprocal and backcross
dumpy wings	(dp dp)	x	wild type (++) reciprocal and backcross
vestigial wings	(vg vg)	x	wild type (++) reciprocal and backcross

4 *Dihybrid cross*
dumpy wing, ebony body (dpdp,ee) x wild type (++,++), reciprocal and backcross
vestigial wing, ebony body (vgvg,ee) x wild type (++,++), reciprocal and backcross
Details of a suitable programme for both these types of crosses are given in Table 15, page 96; similar programmes may be constructed for the following crosses.

5 *Autosomal linkage*
Tight linkage:
yellow body (yy,++) x white eye (++,ww)
Loose linkage and cross-over value in coupling and repulsion:
curled wing, ebony body (cucu,ee) x wild type (++,++) . . . coupling,
curled wing (cucu,++) x ebony body (++,ee) . . . repulsion.
Drosophila behaves very unusually in that there is no crossing over in the male, so that different results are obtained in the backcross depending upon whether the F_1 heterozygous parent is male or female.

6 *Sex linkage*
white eye (ww) x wild type (++)

7 *Population studies, competition within a species*
These are important as illustrating the basis of natural selection, balanced polymorphisms and the Hardy-Weinburg principle.
Three or more population cages (see Fig. 20, page 98) are required. Two contain equal numbers of two pure strains respectively and the others various combinations of these strains totalling initially the same number of flies. In each case the number of males and females is the same. The populations are examined and scored at weekly intervals for a minimum of twelve weeks. Suitable pure strains are the wild type, ebony body and vestigial wing.[14,20]
It is a useful extension of these population studies to determine whether selective matings occur. Drosophila can often be seen posturing before mating and if sample counts are made recording the numbers of each of the possible combinations of pairs observed, useful data can be collected. These are of course in their simplest form in the initial stages of a population when the numbers of the two phenotypes are equal.

8 *Salivary gland giant, or polytene, chromosome preparation*
The banded structure of these bundles of homologous chromosomes is evidence of a linear structure in the chromosomes and hence evidence that genetic factors may have physical counterparts (genes) situated linearly along the chromosomes. The glands are taken from fully grown larvae just as they begin to crawl up the sides of the container to pupate.[7,33,75]

Preparation of polytene chromosomes: If larvae are required for this purpose they should be raised at a low temperature (15°C) and in uncrowded conditions.

a) Hold the larva firmly towards its rear with a dissecting needle. Then, with a downward and forward movement with the side of the point of another dissecting needle, separate the head end from the rest of the body and draw it away. (Alternatively the rear of the larva may be cut off and the body contents pulled out by detaching the head using a dissecting needle.) The glands will generally be drawn out cleanly with the head whilst decapitation kills the animal instantly. They can be recognised as two flattish structures attached to the anterior of the gut with large cells and prominent nuclei.
b) Detach the glands and transfer them immediately into a drop of acetic orcein stain solution on a

slide. (On no account must they be allowed to dry and it may be advantageous to dissect them out under saline solution). Leave to stain for five to ten minutes. It is not usually necessary to heat.

c) Apply a cover slip and flatten the preparation by applying slight pressure under filter or blotting paper.

d) Examine under a microscope. If a permanent preparation is desired, then it is necessary to film the underside of the coverslip with glycerine albumen and to proceed as described on page 48.

Flour beetles, Tribolium spp. (2n = 20 or 18) *4, 11*

Two species, *T. castaneum* and *T. confusum*, of this genus are being used increasingly in genetical studies, especially in connection with the storage problems of cereal products and similar dry foods in which these beetles are considerable pests. For school studies their uses are limited and they are not a serious alternative to *Drosophila.*[12,46,47,63]

Mutants Suitable for School Studies

In *T. confusum* there are nine linkage groups and a large number of mutants have been recognised but only three have proved really suitable for school work, these are listed in Table 8.

Suggested investigations

1 *Monohybrid cross*

pearl eye (pp) x wild type (PP), reciprocal and backcross

black body (bb) x wild type (BB), reciprocal and backcross

F_1 all bronze due to semi-dominance

F_2 1 black: 2 bronze (heterozygotes): 1 wild

2 *Dihybrid cross*

pearl eye wild colour body (pp,BB) x wild eye colour black body (PP,bb)

F_1 pearl eye and black body

F_2

Eye colour	Body colour	Genotype	Numbers	
black	brown	PP,BB	3	9
black	bronze	Pp,Bb	6	
black	black	Pp,bb		3
pearl	brown	pp,BB	1	3
pearl	bronze	pp,Bb	2	
pearl	black	pp,bb		1

Table 8 Tribolium confusum *mutants suitable for school studies*

Linkage group	Name	Symbol	Wild type	Mutant type	Suggested use
II	pearl eye	p	Black eye	pale, almost colourless eye.	monohybrid, modified dihybrid.
	ebony body	e_2	Brown body	black, shiny body	monohybrid.
III	black body	b (semi dominant)	Brown body	black, dull body	monohybrid, modified dihybrid

3 *Population studies*

This is probably the area of study in which *Tribolium* will prove most useful at the school level since the cultures need very little attention and can be left over school holidays. The pattern will be much the same as described for *Drosophila* (see page 96). Interesting results have been obtained when mixed populations of the same initial proportions have been kept at two different temperatures, for example, 30°C and room temperature.

Grasshoppers and locusts

These Orthopteran species are taken together because they can be used for the same purpose, namely the preparation of testis squashes for the study of meiosis, but they have advantages and disadvantages so that a choice between them will have to be made in the light of the circumstances under which the work is carried out. The males are heterogametic with only one X chromosome.[28,43,53,69]

Chorthippus spp. (short-horned grasshoppers) (2n = 17 ♂, 2n = 18 ♀) *8*

These have the advantage of having fewer and larger and more distinctive chromosomes than either of the locusts. In addition they have some chromosomes with terminal (or nearly terminal) centromeres and others which are mediocentric so that the difference between the first and second meiotic division is immediately apparent.

Removal of the testes. The insects are best killed by cutting off their heads. The testes are then removed by squeezing the contents of the abdomen on to a microscope slide and teasing away the investing fatty material. The testes are fairly compact masses of tapering follicles surrounded by the bright yellow fat body. The testes can be fixed in acetic alcohol (at least 15 minutes) and stored in 70% alcohol in a refrigerator for several months without deterioration so that the seasonal nature of the material can be overcome. The cells in a follicle show a progression of the meiotic stages, so that if a follicle can be removed singly and squashed gently without displacing the component cells too violently the succession of stages can be made out.

Locusta migratoria (migratory locust, 2n = 23 ♂, 2n = 24 ♀) 8 and Schistocerca gregaria (desert locust, 2n = 23♂, 2n = 24 ♀) 8

The great advantage of locusts for meiosis preparations is their ready availability at all times of the year. Many schools, perhaps most schools, now keep them for a number of purposes. Their drawback lies in the greater difficulty in making out the finer detail and in interpreting the relationships in the preparations made with locust material as compared with grasshopper material. The decision as to which to choose probably rests on how much analysis of the metiotic figures is to be attempted. Most schools make the preparations only to illustrate the general method and rely upon prepared slides or even photographs for the more critical work. For this approach the locust probably offers the better alternative. It is a pity, however, especially at sixth form level, not to get as much value as possible from this particular study.[28,43]

Removal of the testes. Removal of the testes can be carried out as described for the grasshopper or by dissection. For the latter the abdomen is opened up by a longitudinal cut along the mid-line of the dorsal surface of a male hopper. The testes have a form resembling a bunch of bananas and are whitish in colour surrounded by the bright orange fat body. Follicles can be teased out from the surrounding fat by needles and fixed, as described for *Chorthippus*, alternatively they may be used fresh.

Squash preparation of testis material

a) Carefully tease a small portion of the material in a drop of acetic-orcein stain solution (acetic-lacmoid or acetic-carmine could be used as an alternative).
b) Gently heat until the solution steams, but without boiling, two or three times. It is preferable to use a spirit flame. The amount of heating should be varied according to the hardness of the tissue. Add more stain solution to ensure that the preparation does not dry.
c) Leave for about ten minutes. It may then be necessary to heat again and/or to tease the tissues again.
d) Place a coverslip on the preparation. (If a permanent preparation is to be made the undersurface of the coverslip should be filmed thinly with glycerine-albumen).

e) Place several thicknesses of filter or blotting paper over the coverslip and apply pressure to squash the preparation. Do not move the coverslip sideways or the cells will be piled on top of each other.
f) Examine under low and high power of the microscope.
g) To complete a permanent preparation:
   i) Separate the coverslip from the slide by immersing the slide in acetic alcohol when acetic-orcein or acetic-lacmoid is used. (Immerse the slide in 10% acetic acid if acetic-carmine is used).
   ii) Pass the slide and coverslip through two or three changes of absolute alcohol allowing about two minutes immersion each time when acetic orcein or acetic lacmoid is used. (If acetic-carmine is used the slide and coverslip must first be immersed in acetic-alcohol for about two minutes).
   iii) Mount in 'Euparal'. If Canada Balsam or other similar mountants are used then it is necessary to rinse the slide and coverslip in xylene before mounting.

Further details of the removal of the testes and staining may be found in references 1, 7, 12, 14 and 20.

5 Flowering plants– greenhouse plants

So far we have been concerned with animals; the remaining chapters deal with plants and micro-organisms, so that it will be useful at this point to make a comparison between plants and animals as material for teaching genetics. Since the basis of inheritance is fundamentally the same in all forms of life, as wide a range of organisms as possible should be used in order to emphasise this. In many ways animals are easier to work with, firstly because the two sexes are usually separate so that there is no problem of emasculation, secondly the life cycles of many are very short and can usually be made independent of the time of the year, and thirdly mating between the sexes normally takes place once the two sexes are put together. On the other hand animals need more attention, are generally more costly and the progenies may be smaller.

Two main examples are used to illustrate the genetical uses of greenhouse plants. The first is the tomato (*Lycopersicon esculentum*) usually grown from seeds and the second the so-called geraniums (*Pelargonium spp.*) which are normally propagated from cuttings but which can be grown from seeds.

Management
See Table 17, page 101.

Tomato, Lycopersicon esculentum (4x = 24) *1, 2, 4, 12, 13, 14, 17*

The tomato was introduced into Europe from Mexico, where it had been cultivated for centuries, soon after the conquest of that country by the Spaniards early in the sixteenth century. Its original home is thought, however, to have been in Peru or Equador.

The tomato has been found exceptionally useful for the teaching of genetics; a number of mutants are now readily available for demonstrating most of the important genetical principles. It should be stressed however that much of the value to be gained from genetical investigations lies in following through all the breeding processes as part of an enquiry and not merely in recording the results obtained from prepared crosses, valuable though they may be.

Above 600 mutants in the tomato have now been identified or artificially induced. Rapid progress is being made in mapping the twelve chromosomes. For school purposes when greenhouse facilities are limited, it is obviously best to use characters that are recognisable in an early seedling stage. This naturally makes it more difficult to find material for experiments on linkage, which are perhaps better left to animal material, although it is hoped that suitable loosely linked mutants in tomato may become available before long.

Information about mutants that have been found suitable for school work is given in Table 9.

Suggested investigations

1 *Variation*
Seeds known to be carrying mutations in the homozygous condition or to be segregating for them can be sown and selected plants grown to maturity. The range of the variations can be studied and those characters considered most useful for breeding experiments picked out. Pleiotropic effects may be seen, for example, in plants homozygous for Baby Lea syndrome.

2 *Interaction between genetic constitution and environmental factors* (see page 62 for similar demonstrations with barley)
The expression of the anthocyanin factor in the wild type plant, namely the purple stem and deeper colouring of the leaves, is greatly affected by temperature, light intensity and nutrient status, the colour being enhanced by low temperature (10°C as against 20°C), high illumination and low fertility. Seedlings known to be segregating for Baby Lea syndrome can be grown in various combinations of these conditions, or simply low

Table 9 Tomato mutants suitable for school studies

Number of Chromosome	Name	Symbol	Plant Part Affected	Descriptive Notes
1	colourless fruit epidermis	y	Fruit	In homozygous condition (yy) the fruit appears pink due to the red carotenoid flesh under the colourless skin. Recessive.
2	anthocyaninless	a	Whole plant	No anthocyanin can be produced when in homozygous condition. There are a number of mutants (see below) affecting anthocyanin production presumably by blocking different steps in the biosynthesis of the pigments. Recessive.
	mouse ear	Me	Leaves	Incomplete dominant: the leaves are pinnately compound with reduced segments and the internodes short. Homozygote (MeMe) is viable.
3	Baby Lea syndrome	bls	Whole plant	In the homozygous recessive condition, like in anthocyaninless above, no anthocyanin can be produced. The expression of the dominant purple colour of the wild type plant is sensitive to environmental conditions and the action is pleiotropic. First discovered in the variety Baby Lea. Recessive.
	Curl	Cu	Leaves	The leaves reduced and distorted when gene is present and they clasp the stem and give it a very strange appearance CuCu is viable. Dominant.
4	entire leaf	e	Leaves	First leaf is undivided, later leaves are wider than wild type and almost entire. Can be distinguished from potato leaf by indentations. Recessive.
6	potato leaf	c	Leaves	First leaf is almost undivided, later leaves much less divided than wild type, leaflets almost entire and with no indentations. Recessive.
	thiamineless	th	Whole plant	Lethal recessive in homozygous condition; seedlings pale, spotted and distorted but can be restored by foliar sprays of thiamine. Recessive.

7	green stripe	gs	Fruit	In homozygous condition unripe fruit have vertical green stripes, ripe fruits vary in appearance according to other genes present. Recessive.
8	green flesh	gf	Fruit	In homozygous condition the chlorophyll does not disappear from the fruit as it ripens so that flesh appears brown. Anthers have a green tone. Recessive.
9	Hoffman's anthocyaninless	Ah	Whole plant	The recessive allele in the homozygous condition, like the other two anthocyanin mutants mentioned above, gives green stem in seedlings. Dominant.
	pumila	pum	Whole plant	Plumule develops slowly in the recessive homozygote resulting in very small plants. Recessive.
10	gold or Xantha-2	xa-2	Whole plant	A chlorophyll mutant which is lethal in the homozygous condition is expressed as a golden colour in the heterozygote, which can be brought to maturity. Incomplete recessive.
	tangerine	t	Flower and fruit	In the homozygous condition the anthers and the fruits have a distinctive tangerine colour. Recessive.
11	hairless	hl	Whole plant	The homozygous recessive is completely hairless; plants tend to be weak and brittle. Recessive.

temperature high illumination and high temperature low illumination, and the results studied. The Baby Lea plants serve as a control.

3 *Monohybrid crosses*

3.1 *Segregation showing dominance*
anthocyaninless (aa) x wild type (++)
Baby Lea syndrome (blsbls) x wild type (++)
potato leaf (cc) x wild type (++)
thiamineless (thth) x wild type (++)
pumilla (pumpum) x wild type (++)
hairless (hlhl) x wild type (++)

3.2 *Segregation without dominance*
Gold or Xantha-2. The material usually supplied segregates one normal green, two golden heterozygotes, one white homozygous recessive. The white seedlings soon die but with care the heterozygotes can be brought to maturity and will, of course, when selfed provide seed for the following year. Although tomatoes are normally self pollinated it is wise to isolate the plant or plants being grown to maturity for this purpose so that they cannot be contaminated by foreign pollen.

4 *Dihybrid crosses*

4.1 *Segregation with dominance for both characters*
anthocyaninless (aa,++) x hairless (++,hlhl)
anthocyaninless (aa,++) x potato leaf (++,cc)
Baby Lea syndrome (blsbls,++) x hairless (++,hlhl)
Baby Lea syndrome (blsbls,++) x potato leaf (++,cc)

4.2 *Segregation giving 9:7 ratio*

P green plant, Hoffmans anthocyaninless (ahah,BlsBls) x green plant, Baby Lea syndrome (AhAh,blsbls)

F_1 purple stemmed plants, heterozygous for both characters (Ahah,Blsbls)

F_2 purple stemmed plants: green stemmed plants in 9:7 ratio.

The synthesis of anthocyanin is blocked at two different points which is evidence supporting the biochemical action of the gene.

5 *Tight linkage*

P Hoffman's anthocyaninless/pumilla: homozygous dominant for both alleles (AhPum/AhPum) x homozygous recessive for both (ahpum/ahpum)

F_1 Hoffman's anthocyaninless/pumilla heterozygous for both alleles (AhPum/ahpum)

The recessive homozygotes for both characters can be recognised in the early seedling stage by the absence of purple colour in the stem (ahah) and by the characteristic hooked appearance of the first true leaf (pumpum) so that scoring for an F_2 or backcross generation can be done at the seedling stage which saves a further year of work

Backcross to the double recessive parent:

AhPum/ahpum x ahpum/ahpum gives AhPum/ahpum: ahpum/ahpum in 1:1 ratio instead of the 1:1:1:1 ratio, Hoffman's anthocyaninless pumilla: Hoffman's anthocyaninless normal leaf: normal anthocyanin pumilla: normal anthocyanin normal leaf that would be expected with independent assortment.

The recombination value is in fact about 5% and if a large number of seedlings can be grown this value can be determined. This is a case where cooperation between a group of schools would be beneficial.

The demonstration is more convincing if plants homozygous for the two genes separately are grown at the same time or, better still, if each is crossed with the wild type and shown to segregate normally.

6 *Biochemical supplementation*

This is a very important demonstration since it provides strong evidence for the biochemical action of a gene in a higher organism, thus supporting the evidence derived from the study of many nutritional mutants in bacteria and fungi. The homozygous recessive of thiamineless (tltl) dies a few weeks after germination although it can be brought to maturity by foliar sprays of thiamine (Vitamin B_1) two or three times a week throughout the life of the plant. The heterozygote (tlTl) can be recognised early on by a slight chlorosis if the plants of an F_2 progeny are grown under optimum cultural conditions. The mutant can be maintained by two methods (see page 104).

Two steps are necessary to provide the evidence required to show biochemical supplementation. First, as suggested above, it must be shown that the condition is due to a single gene action. Second it is necessary to show that the condition is due to the inability of the mutant to form thiamine by correcting the deficiency by supplying thiamine. If the supply is stopped the severe chlorotic condition returns.

Thiamineless seedlings respond to pyrimidine sprays as well as thiamine which shows that the block occurs at the point marked tl/ in Fig. 3 showing the synthesis of thiamine.

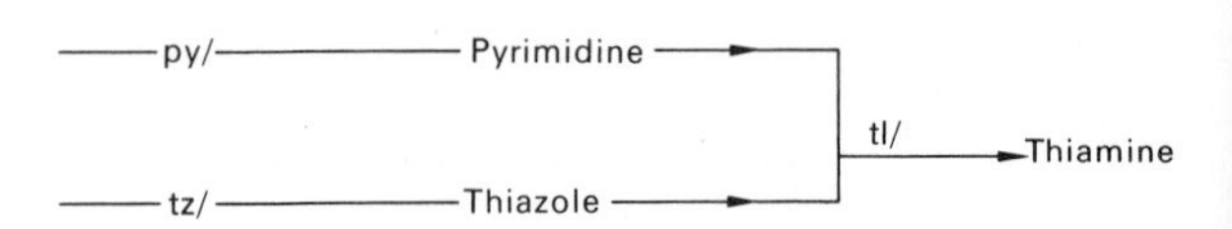

Figure 3 Biosynthesis of thiamine

Geraniums, Pelargonium spp. (X = 8, 9, 10 or 11)*1, 4, 9, 10*

The familiar geranium is one of the easiest pot plants to propagate and grow as it can stand considerable neglect and withstand a wide range of environmental conditions. 'Geraniums' as we know them, although members of the natural order Geraniaceae do not belong to the genus *Geranium* but to another related genus *Pelargonium*, the members of which are principally native to South Africa where they are often exposed to harsh conditions and long periods of drought. The house and bedding-out 'geraniums' have a very mixed ancestry, the commonest group, the zonal 'geraniums' having, in all probability, been derived by hybridisation during the two hundred years or so, since they were introduced to Europe, between at least four species, *P. inquinans, P. zonale, P. graveolens* and *P. scandens.* Zonal geraniums, with which we are chiefly but not exclusively concerned here, are now correctly referred to as *P. hortorum.* Much of the use of these plants will be incidental; they can be brought in to illustrate various points even though their main purpose is a decorative one. The names of suitable cultivars are given as suggestions.[86]

Suggested demonstrations and investigations

1 *Variation*

1.1 *The range of variation to be found within a genus and demonstration of the use of hybridisation in the development of decorative plants*
If the species types can be obtained for comparison so much the better, but it must be emphasised that the ancestral species given are hypothetical and relate only to our present stage of knowledge.
Regal Pelargoniums derived from *P. domesticum:*
Aztec (white, strawberry and brown petals; colour varies with fertilizer status), Caprice (deep pink), Blythwood (deep mauve), Conspicuous (wine red), Lord Bute (purple, picotee edge).
Zonale Pelargoniums derived from *P. inquinans* (gave colour); *P. zonale* (gave leaf zone), probably *P. graveolens* and *P. scandens* and perhaps others, see 1.2 below.
Sweet Scented Pelargoniums derived from *P. crispum, P. graveolens* and *P. fragrans:*
Fragrans (very close to *P. fragrans*), Royal Oak (oak shaped leaves), Attar of Roses (rose scented), Little Gem (lemon scented, deeply cut leaves), Prince of Orange (orange scented), Shottesham Pet (nut scented).
Unique Pelargoniums derived from *P. fulgidum:*
Purple Unique (a good representative)
Ivy-leaved Pelargoniums derived from *P. peltatum:*
Peltatum (very close to the species), Galilee (double, rose pink), Red Galilee (double red mutant from Galilee), Snowdrift (double, white flowers, pale leaves), Mexican Beauty (single red), Madame Crousse (single pink), L'Elegante (single white with purple veins).

1.2 *Variation in flower form and colour, leaf shape and variegations, plant size and habit*
The following are easily obtainable varieties from the zonale group:
Single: Snowstorm (white), Beatrix Little (scarlet, small plant), Doris Moore (red), George Burg (violet-purple semi-primitive).
Double: Snowball (white), Mrs Lawrance (pink), Paul Crampel (red), Paul Grozy (red, small).
Variegated leaves (Chimaeras): Happy Thought (green outside white), Crystal Palace Gem (yellow outside green), Flower of Spring (white outside green), New Life (two colour layers in the flower)
Graft Hybrid: Skelly's Pride (*P. peltatum* outside, *P. zonale* inside), Mrs Henry Cox (complex).

2 *Demonstration of a haploid plant and its diploid derivative*
A miniature geranium Kleine Liebling which is much prized for its dainty form and good flower colour but which never produced seed was found on examination to be haploid. Haploid plants are known to occur very infrequently in nature and a few have been produced by tissue culture but here is one that can be easily grown and propagated. Not only that, a normal diploid has been produced by mutagen treatment so that the two can be grown as a demonstration side by side.

3 *Chlorophyll mutant segregating 1:2:1 in the* F_2
The varieties Golden Crampel and Golden Crest, which have pale gold leaves, are heterozygous for a white lethal mutant. If the flowers are enclosed in cellophane bags in the bud stage and subsequently self-pollinated or better still pollinated with pollen from another plant of the same variety, the resulting seeds will give normal, golden and white seedlings in the ratio of 1:2:1. The white seedlings, being completely devoid of chlorophyll, soon die but the golden seedlings can be brought to maturity (compare with the Xa-2 heterozygote in tomato). Each flower only produces five seeds so that the number of seedlings available for scoring is likely to be small but this will give an opportunity for discussing the tests of significance appropriate when numbers are limited. The beautiful parachute mechanism of the seeds that in nature assists in dispersal is a feature that makes the ripening of the flower heads worthwhile.

Primulas *4, 9*

Some work with hardy *Primula* species is described in Chapter 7; several greenhouse species offer opportunities for complementary studies.

Suggested demonstrations

1 *The origin of Primula kewensis, (4x = 36) 9*
P. kewensis is the bright yellow flower which makes such a fine display in florists' shops at Christmas time. It arose as a spontaneous hybrid in 1897 just at the time when scientific interest in genetics was being aroused and its study led to new understanding about how new species could be formed and hence the process of evolution. The story of this historic plant is worth giving in some detail before we discuss how best to arrange the greenhouse demonstration.[8 7]

The original seedling was found in one of the greenhouses at the Royal Botanic Gardens at Kew. It was clear, because of certain similarities and from an examination of the house for possible parents that it was a chance hybrid between two species, *P. verticillata* and *P. floribunda.* Two points are worth noting, first that *P. verticillata* was a native of Arabia and *P. floribunda* of Afghanistan so that until then the two parental species had been geographically isolated with no opportunity for hybridisation, second that the hybrid, rather unusually, inherited the best features from a commercial point of view from both its parents. Thus it had the erect habit and large flowers of *P. verticillata* and the larger number of flowers (as its name suggests) and the more attractive leaf shape of *P. floribunda.* It did however have the white powdery 'farina' on its leaves of *P. verticillata* which some people think detracts from its appearance. Quite obviously its commercial success was assured but like so many hybrids, it did not set seed so it had to be propagated vegetatively, which fortunately was very easy. Then, at least twice a few years later, plants did set fertile seed and the plants could from then on be propagated in that much more convenient way. Chromosome studies of the root tips subsequently confirmed that these new plants were allotetraploids with a complete set of chromosomes 2n = 18 from each parent and hence capable of normal meiotic division. But that is not the end of the story. Later a mutation to shiny green leaf instead of the mealy leaf occurred and the seed sold nowadays is almost invariably of this form so that it is not easy to obtain the original form, though fortunately stocks are being built up for demonstration purposes. To complete the story a variety of *P. floribunda* called Isabellina with pale yellow petals arose as a sport so a very interesting demonstration can be laid out as in Fig. 4.

2 *Studies with P. sinenis 1, 4*
This is another favourite greenhouse perennial. The flowers have two distinct forms, fimbriata, or fringed, and stellata, or star-shaped, and in each there is a wide range of flower colour and leaf shape so the variation can be shown quite dramatically. Equally, a cross between say a white fimbriata and a red stellata will demonstrate a phenomenon plant breeders are familiar with namely a uniform F_1 (hence the

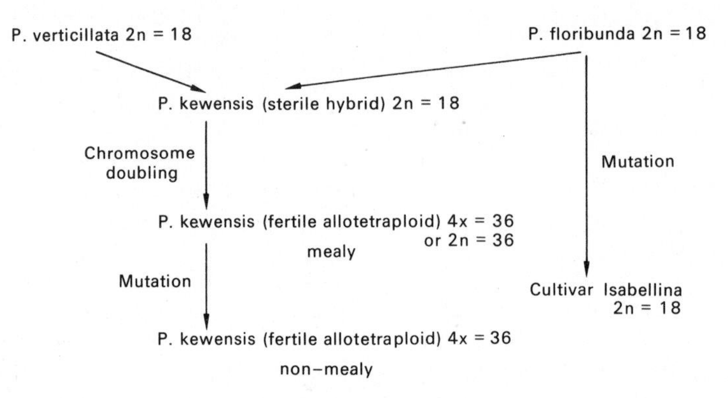

Figure 4 The origin of Primula kewensis

modern F_1 hybrids) and a very variable F_2 which quite defies analysis, the result of many genes being involved.

3 *Studies with P. obconica and P. malacoides 4*
These are two other popular greenhouse plants the first a perennial and the second an annual. Both lend themselves to breeding experiments with flower colour. The genetic basis of heterostyly and pollen forms can also be the subjects for projects.

Root tips: cut off about a 2 mm length of the tip for use. It may be necessary to gently heat the tip in hydrochloric acid before staining to facilitate the separation of the cells. The root tips may be pre-treated in colchicine, which however is very expensive or paradichlorobenzene solution which inhibits spindle formation and thus gives good metaphase plates.

Rhoeo discolor, (2n = 12) *8*

This well-known greenhouse plant is worth growing especially for cytological studies in meiosis and haploid mitosis of pollen grains. Plants can be grown from seed or purchased as mature house plants. The value of this plant for chromosome studies lies in the fact that the chromosome number is small and the individual chromosomes large and distinguishable. In addition, abnormal translocations involving every chromosome are common and these lead to inviable pollen grains which can be distinguished from viable normal pollen by staining in aniline blue. The preparations are made in essentially the same way as for locust testis squashes.[7,9] 8

Squash preparation of pollen mother cells (or root tips)
The procedure is almost identical to that described on page 47; only those points of difference are noted here.

Pollen mother cells: select buds in which the anthers are almost transparent. Stain in acetic-orcein. Remove all visible pieces of debris before adding the coverslip.

Cucumber, Cucumis sativus (2n = 14) allotetraploid forms (4X = 28) are also known *4*

Cucumber plants are monoecious with separate male and female flowers and with different proportions of the two flower forms in different sections of the long trailing stem, an adaptation encouraging cross pollination. Fairly recently seeds have become available for monohybrid breeding investigations involving a substance, cucurbitacin. This substance is normally present in cucumber plants, to which it gives a bitter taste which however, is usually, but not always, absent from the fruit. The recessive mutant nb (non bitter) gives the whole plant the typical cucumber non-bitter taste and is being bred into modern varieties.

Monohybrid cross: normal (Nb) x non-bitter (nb)

The seedlings raised in the F_2 can be tested for the presence or absence of bitter taste by removing very small (2 mm) pieces from the cotyledons and applying them to the back of the tongue where the taste buds for bitter taste are most concentrated. The planning of the tasting procedures to avoid errors due to fatigue etc. and the arranging for checks can be a valuable part of this experiment.

6 Flowering plants– half-hardy annuals

Management
See Table 17, page 101.

Snapdragon, Antirrhinum majus (2n = 16). *1, 2, 4, 10, 14, 20*

The garden antirrhinum is a native of the Mediterranean region though it can often be found in something very close to its wild form naturalised in Britain, generally on old walls. In its native home and when naturalised it is perennial in habit but it is usually treated in Britain as a half-hardy annual. As the flowers are large and very easily manipulated for breeding investigations, antirrhinums provide excellent opportunities for genetical studies of several kinds (see Table 10).

Suggested investigations

1 *Variation under domestication*
Garden antirrhinums can be obtained in a wide range of forms and make a fine display of colour and form in a school garden. The demonstration can be made very spectacular if a range of modern forms are grown close to a few plants of the wild form.

The following cultivars are suitable:
Range of colour in intermediate height for example Sutton's Triumph Strain in White, Primrose, Yellow, Orange, Scarlet, Pink, Crimson, Mauve.
Range in height, same colour for example scarlet or pink, Dwarf Bedder (20 cm), Intermediate (35 cm), Tall (80 cm). (The range in height has been achieved by hybridisation with a dwarf species *A. nanum*)
Different flower form, for example, normal, double flowers (Madame Butterfly), peloric or bell-flowered (Super Pink)
Diploid and tetraploid forms for example Tetrasnaps.
A variety in which the flowers are striped in a variety of colours due to an unstable colour gene (an example of what is known as a colour break), Bizarre.

If a small bed of an F_1 hybrid, which will be very uniform in every respect, is grown alongside a bed of plants raised from seed saved from a similar bed of the F_1 hybrid grown the previous year and allowed to interpollinate, the nature of independent assortment is emphasised since there is a complete break down both of colour and height, due to the recessive genes carried in the interpollinated hybrid.

Table 10 Antirrhinum *varieties—phenotypes, genotypes and details of the pigments they possess*

Variety	Phenotype	Aureusidin	Cyanidin	Pelargonidin	Genotype
White	Pure white	–	–	–	rr, B–, I–
Bright Yellow	Yellow	√	–	–	rr, B–, I_A–
Royal Rose	Pink	–	–	√	R–, bb, I–
Scarlet Triumph	Scarlet	√	–	√	R–, bb, I_A–
Mauve Beauty	Mauve	–	√	–	R–, B–, I–
Rich Crimson	Crimson	√	√	–	R–, B–, I_A–

2 *Inheritance of resistance to 'rust' (Puccinia antirrhini)*
Suitable varieties can be chosen from almost any seed catalogue and the following have been found suitable:
Sutton's Crimson Monarch: crimson, rust resistant, intermediate height (40 cm)
Sutton's Eclipse: crimson, rust susceptible, intermediate height (40 cm)
Carter's Crimson Dwarf: crimson, rust susceptible, dwarf (20 cm)

These can be used in combinations for monohybrid and dihybrid crosses and backcrosses.

In most seasons rust will appear in early autumn and all too frequently it will completely ruin a garden display. Nevertheless for the purpose of carrying out experiments on rust resistance it is desirable to be able to infect the plants at some chosen time.

To do this leaves from badly infected plants should be collected in autumn, dried off and then kept in a plastic bag. When required in the following year the infected leaves should be stirred in a jar of water with a little detergent added so that the dormant spores are well distributed in the water which can then be sprayed on to the growing plants with a fine syringe. The best time to do this is fairly late on a cool evening. Rust will begin to appear in susceptible plants after about two weeks and counts can be made two or three weeks later. It is good technique to test the procedure by spraying control plants of a very susceptible variety such as Malmaison, but if Eclipse, which is also very susceptible, is used, this is hardly necessary.

3 *Inheritance of flower colours*
The antirrhinum is still often quoted in text-books as giving an intermediate pink flowered F_1 when plants bearing red flowers are crossed with plants bearing white flowers. Unfortunately the strain with pure white flowers which was used in the original experiments is no longer available and the inheritance of flower colour is not simple, but it has been intensively studied over a long period of time and it can be used at school level to demonstrate gene interaction which holds the key to gene action. In fact as long ago as 1907, Miss Wheldale, who carried out the original work on the crossing of red and white antirrhinums at the John Innes Horticultural Institute, predicted that studies of this nature would throw light on the biochemical aspects of gene action. Since then the chemistry of flower pigments has gone a long way and simple techniques have been developed for isolating and identifying the pigments to be found in any particular variety.[85,89,103]

The complexity of the situation is indicated by the fact that two anthocyanins, cyanidin and pelargonidin, are responsible for the 'red' colours and no fewer than seven flavones for the 'yellows'. In addition there are genes that affect the distribution of the pigments in different parts of the flower.

For our purpose we can concentrate on the effects of three genes:

a) The dominant gene R without the presence of which no anthocyanin can be formed. This provides an example of epistasis.
b) The dominant gene B which in the presence of R is responsible for the production of cyanidin, which is magenta in colour. In the presence of R but the absence of B, pelargonidin, which is pink, is produced.
c) The dominant gene I_A is responsible for the production of one or both of the flavones aureusidin 1 and aureusidin 2.

Details of varieties suitable for demonstrations are given in Table 10.

Extraction of pigment

a) Crush the upper lips of 20 flowers in $10cm^3$ acid alcohol.
b) Heat the mixture gently in a water bath. Strain or filter the solution.
c) Store the extract in dark coloured stoppered bottles in a refrigerator. Under these circumstances this can be kept for up to a year without deterioration.

Separation and identification of pigments

a) 'Spot' the extract on to a strip of chromatographic paper (or filter paper) about 1 cm from the end using a fine pipette. Apply ten times allowing drying between application. Keep the diameter of the 'spots' less than 0.5 cm.[25]
b) Separate the pigments by supporting the paper strip so that the 'spotted' end is just immersed in a butanol, acetic acid, water solvent (6:1:2) mixture. Allow the solvent to move to the supported end of the strip.

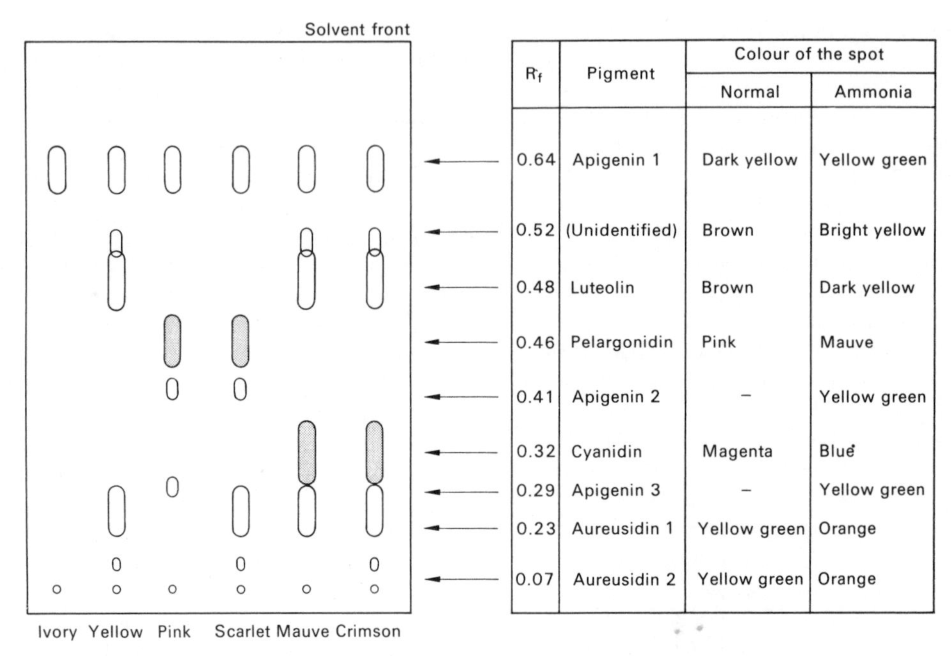

R_f	Pigment	Colour of the spot	
		Normal	Ammonia
0.64	Apigenin 1	Dark yellow	Yellow green
0.52	(Unidentified)	Brown	Bright yellow
0.48	Luteolin	Brown	Dark yellow
0.46	Pelargonidin	Pink	Mauve
0.41	Apigenin 2	–	Yellow green
0.32	Cyanidin	Magenta	Blue
0.29	Apigenin 3	–	Yellow green
0.23	Aureusidin 1	Yellow green	Orange
0.07	Aureusidin 2	Yellow green	Orange

Figure 5 Chromatogram and table of pigments of Antirrhinum.

c) Dry. Note colours. Expose the strip to ammonia fumes. Note changes in colour and the positions of the colour patches.

The relationships of the colour types suggested for the breeding experiments are as shown in Fig. 5.[17]

It will be noted that apigenin 1 occurs in all six colour types. It can be argued that it was absent in the pure white with which the original work giving the clear cut intermediate pink F_1 was done and that it is the presence of this pigment that is responsible for the confused results that are obtained with modern white types. Apigenin 2 always occurs when pelargonidin and the two aureusidins are found together. It is on the basis of these observations that the three major genes R,B and I_A referred to in the previous section have been postulated.

Maize, Zea mays (2n = 20) *4, 14, 16*

Maize is a tropical or sub-tropical crop plant of great antiquity in the western hemisphere and now of great economic importance throughout the world. Maize, in its present form, could not compete in the wild and no ancestral wild form is known. It is therefore a plant of considerable archaeological and botanical interest. It has also contributed much to genetical theory. It is an admirable subject for genetical studies in schools in places where it can be grown, which in Britain means the southern part of England, although quite useful results can be obtained from plants grown in pots in greenhouses further north.

Several features make maize a suitable plant for genetical studies:

a) It is wind pollinated and large male and female inflorescences occur widely separated on the same plant. There are therefore no difficulties in emasculation and pollination.
b) Each plant will produce at least two "cobs" each bearing up to 500 kernels which, when kernel characters are used, make attractive and valuable museum specimens.
c) Maize has been very extensively studied; many mutants are available to illustrate various genetical principles and the chromosomes have been thoroughly mapped.[94]

Suggested investigations

1 *Monohybrid, dihybrid and trihybrid crosses*
As stated, time is gained if 'seed' characters are used in plant breeding experiments. In maize three characters affecting the colour or the appearance of the endosperm of the kernel have proved particularly useful in elementary breeding studies although others are also available. A point concerning the endosperm has to be borne in mind although it does not affect the results if suitable characters are chosen. This is that the endosperm is derived from the fusion of the vegetative nucleus of the pollen tube and diploid cell of the embryo sac so that it is triploid in nature having two sets of chromosomes from the female parent. This only appears to affect the degree of the expression of the character in the characters chosen here.

The three factors are:

Aleurone colour, C, the presence of which allows colour to be formed in the alenrone layer of the endosperm. If absent no colour is formed even if the genes for pigmentation are present. C is therefore an epistatic factor.

Red aleurone colour, pr, a recessive gene which if present in the homozygous condition gives red instead of purple aleurone colour.

Sugary endosperm, su, another recessive gene which if present in the homozygous condition, results in an endosperm which is translucent and wrinkled at maturity owing to the nature of the storage substances.
It follows that

PrPr,CC	PrPr,Cc	Prpr,CC	Prpr,Cc	give purple grains
prpr,CC	prpr,Cc	give red grains		
and	PrPr,cc	Prpr,cc	prpr,cc	give white or yellowish grains

These factors enable monohybrid (red x white), dihybrid (red starchy x white sugary) and trihybrid (purple starchy x white sugary) crosses to be made although, owing to the epistatic factor, the ratios in the three factor cross will be 27:9:9:3:12:4 and 1:1:1:1:2:2 instead of those normally expected.

Since the cobs make good museum specimens it is wise to make one cross and the corresponding backcross each year and thus avoid the problem of contaminating pollen, taking the three crosses in rotation (see page 103).

Once the original parental crosses have been made and F_1 'seeds' obtained the demonstrations can be made self-perpetuating in the following manner. The description given is for the monohybrid cross, but a similar procedure can be used for dihybrid and trihybrid crosses.

Rows of the homozygous recessive plants are sown or planted alternately with rows of the heterozygous F_1 plants spacing the plants about 60 cm apart each way:

Red F_1 row	White recessive row	Red F_1 row
prpr,Cc	prpr,cc	prpr,CC etc.

The male or staminate inflorescences of the 'white' recessive rows are pulled out as soon as they appear so that the only pollen available is that from the 'red' F_1 plants which will be of two kinds in theoretically equal quantities, carrying or not carrying the colour factor C respectively.

This pollen falling on the stigmas of the 'red' heterozygous plants will give the 3:1 F_2 ratio and on those of the 'white' homozygous plants will give the 1:1 ratio for the backcross to the recessive parent. The red grains on the cobs of the 'white' recessive rows of plants are heterozygous and provide the seed for a subsequent year.

A very useful further investigation is to determine the genotype nature of the 'seeds' expressing the dominant character in the F_2 generation in the case of the monohybrid cross. To do this about 50–100 red 'seeds' from a cob showing the 3:1 ratio should be planted in a block and the mature plants allowed to interpollinate freely. We know that the plants are of two kinds, those homozygous for aleurone colour (CC) and those that are heterozygous (Cc). The pollen will be of two kinds, carrying C and carrying c. It is a useful exercise to ask the class the theoretical ratio for the two kinds of pollen although the ratio is not material to the method. If all the plants produce roughly the same quantity of pollen whether they are homozygous or heterozygous the ratio is of course two carrying C to one carrying c. The essential point however is that plants which are homozygous produce cobs containing all red grains, whereas heterozygous plants produce cobs bearing some white grains. It is important to appreciate that in this case it is the kinds of plants (homozygous or heterozygous) that is to be counted and it simplifies matters to reduce the number of cobs to one per plant.

2 *Chlorophyll deficient mutants*
In common with many green plants intensively used for genetical studies a number of chlorophyll mutants have been found or induced artificially in maize (see pages 51 and 102). Amongst these is a lethal albino mutant, white seedling (w_2). 'Seeds' segregating 3 green: 1 white seedlings can be obtained from biological suppliers and can be used for physiological studies as well as for demonstrating genetical ratio in the seedling stage. These seeds can be quite useful in demonstrating the monohybrid F_2 ratio with a minimum of trouble. It is more instructive however to ask how such seeds are obtained and perhaps to maintain one's own stock. The process is simple enough where the plant is normally self pollinated as in the case of barley (see page 102) or where the heterozygous plants are recognisable and can be brought to maturity as in the case of the gold mutant in tomato (see page 51) but it is more difficult in a naturally out-breeding plant like maize. Two methods are possible, controlled self pollination of individual plants and the use of a closely linked marker gene, viz plant colour R.

3 *Dwarf plants and growth-promoting substances*
A number of dwarfing genes have been discovered in maize and these provide useful material for the study of growth promoting substances such as gibberellins and indole-acetic acid (IAA). The 'seeds' usually supplied yield 3 tall: 1 dwarf plant and the differences are recognisable at an early seedling stage so that the material can be used for demonstrating the monohybrid ratio, but a more important function lies in the fact that having shown that the dwarfing is a genetic effect following the laws of Mendelian inheritance, it is possible by treating the seedlings, and later the growing plants, with a suitable growth promoting substance, to produce a phenotypically normal plant which, however, remains genetically dwarf. This is therefore strong circumstantial evidence that the action of at least some genes is biochemical, which is an important step in the logical development of the study of genetics.[96]

The material most easily available is dwarf plant–I, (d_1) which responds to applications of very dilute solutions of gibberellic acid but not to other common growth promoting substances. Investigations can now be designed to examine the responses of dwarf and normal seedlings to various growth substances. In the seedling stage about 1 cm^3 of aqueous solution, giving about 10 μg gibberellic acid in total, dropped on to the growing point by means of a fine pipette about once a week brings a rapid response and in four weeks the treated dwarf plants are usually indistinguishable from the normal control plants. Similar applications of IAA have no perceptible effect. As the plants grow bigger the solutions can be applied all over the leaves and stems by means of an atomiser and the dwarf plants brought to maturity. If now the treated dwarf plants are isolated from all other maize plants and allowed to interpollinate, seeds will be produced from the phenotypically normal, but genetically dwarf, parents and grown again the following year to show that although the parents were apparently tall they still handed on the dwarf character in the same manner as genetical dwarfs. Sometimes the dwarfs obtained by this procedure are somewhat taller than true genetical dwarfs probably due to some gibberellic acid being carried over in the seed.

4 *Storage of amylopectin: genes and germ cells*
The waxy (wx) mutant stores amylopectin instead of normal starch in its tissues. Amylopectin gives a reddish brown colouration with iodine solution instead of the blue-black colouration of starch. The action of the gene can be recognised in the pollen grains, so that if the pollen from a plant heterozygous for the character is treated with iodine solution and examined under the microscope two kinds of pollen grains in equal numbers are found, those which stain blue-black carrying the dominant (Wx) for normal starch and those which stain brown carrying the recessive (wx) for amylopectin. This is direct evidence that the genes separate and function independently in the germ cells. This was a step that Mendel inferred from his other evidence but here his inference is given direct support.

Stocks, Matthiola incana (2n = 14) *1, 2, 4, 20*

Suggested investigations

1 *Variation*

1.1 *Variation in a genus* For example, *M. incana* (Brompton and Ten-week Stocks) *M.bicornis* (Night-scented Stock) and *M. sinnata* (the native Sea-Stock now becoming rare).
1.2 *The variation within a species under cultivation* For example *M. incana.* The study of any good seed catalogue will suggest suitable cultivars, one of which will be the varieties with double flowers which most people prefer to single flowers.

2 *Inheritance and commercial production of double stocks*
In double flowers the stamens are totally or partially replaced by petals so that they are either sterile or to some degree infertile. In a species grown from seed the gene for double flowers has to be carried in the heterozygote and it is obviously an advantage if these can be recognised in the seedling stage because of some closely linked character which shows up early in the life of the plant. In the Hanson's varieties of ten-week stocks not only has this combination been found but a pollen lethal, which of course segregates 1:1, also linked with the other two genes, reduces the ratio of 'singles' to 'doubles'. The three closely linked alleles are therefore single flowers/double flowers; dark green cotyledons and foliage/light green foliage and cotyledons; normal pollen/lethal pollen. Not only that, the difference between dark green and light green seedling colour is expressed most clearly in seedlings germinated at about 15°C but then reduced to about 9°C, so that there is an environmental effect. Although the seeds, which can be easily obtained commercially, are usually used to demonstrate an unexpected ratio, there is obvious scope here for valuable project work.

Other half-hardy annuals suitable for breeding investigations *4*

Morning glory, Ipomea purpurea: Flower colour—for example, Heavenly Blue x Pearly Gates.
Four O'Clock or Marvel of Peru, Mirabilis jalapa: Flower colours—for example Red x White from mixtures.
Both the above need a warm sheltered position.
Tobacco Plant, Nicotiana affinis: Flower colours, height and reaction to daylight—for example Affinis (tall, white, closes by day), Dwarf White (short, white, opens by day), Daylight (tall, white, opens by day), Dwarf Crimson (short, crimson, closes by day). Albino mutants from material segregating 3 green: 1 white.
Petunia hybrida, itself a hybrid between *P. nyctaginiflora* (white) and *P. integrifolia* (purple): Flower colours (self colours and striping; double flowers and ruffled edges)— for example Blue Bee (blue, single), Rose Queen (pink, single), Brass Band (yellow, single), Snowball (white, single), Blue Danube (blue, double), Valentine (red, double), Cherry Tart (pink, double ruffled), Starfire (red and white stripes, single), Starjoy (pink with white star).

7 Flowering plants– hardy field and garden plants, annuals and biennials

Annuals and biennials can be conveniently grown on a small patch of open ground which besides its genetical uses can be used for plot experiments and for demonstrations of cultivations, rotations and various plant treatments.

Management
See Table 17, page 101.

Annuals
Barley, Hordeum spp. (2n = 14) *1, 2, 4, 10*

Suggested investigations and demonstrations

1 *Variation*
All the species in the section Cerealia, which includes all the cultivated species of the genus *Hordeum*, are interfertile. Only one, *H. spontaneum* is unquestionably wild although another, *H. agriochriton*, has the brittle rachis characteristic of wild grasses and was at one time thought to be wild. The most generally accepted view now appears to be that *H. agriochriton* resulted from hybridisation between *H. spontaneum* and a cultivated six-row barley.

Some discontinuous characters of barley and the species exhibiting them are:

H. spontaneum:	two-row, black, awned, brittle rachis
H. agriochriton	six-row, white, awned, brittle rachis.
H. distichum:	two-row, white or black, awned, tough rachis.
H. irregulare:	four-row, black, awned, tough rachis.
H. trifurcatum:	six-row, white or black, trifid awns, tough rachis.
H. gymnospermnum:	two-row, black, hooded awns, tough rachis.
H. hexastichum:	six-row, black, awnless, tough rachis (naked grains).
H. inerme:	two-row, white, awnless, tough rachis.

Black is dominant to white, two-row to six-row, and awned to awnless. The unhulled condition (adherent chaff) is normal although naked-grained varieties do occur.

In addition to the characters mentioned above glumes can be large or normal in size in certain species and varieties.

The toughness of the rachis, so important in the evolution of all the cultivated grasses is a quantitative character though it is difficult to find a scale for toughness. Continuous characters simpler to measure are plant height, ear length, number of grains in one ear, number of grains per unit length of ear (a measure of laxness), number of tillers per plant, total weight of grain per plant, weight of individual grains and so on.

An interesting method of breeding barley varieties adapted to local conditions was to produce what was called a composite cross, which gave seed segregating for a large number of agricultural characters but including all of those listed above. The idea was that, by growing this seed under local conditions and sowing the seed harvested in successive years, strains would eventually be selected out adapted to these local conditions. Mixed seed from the original composite crosses can be obtained from biological suppliers and provides excellent material for the study of variation.

2 *Interaction between genetic constitution and physical environment*
One of the most important concepts that should emerge early in a course in genetics is that, whereas the genetic endowment an individual organism possesses determines its potentialities, the environment determines the extent to which these potentialities are realised.

This idea can be demonstrated quite dramatically by the use of yellow, cholorophyll deficient mutants of green plants. Many of these are known and one of these, a Xantha mutant of barley, is particularly suitable for the purpose.[14]

Two pots are sown with barley seed, one with a normal variety and the other with a Xantha strain known to segregate three green seedlings to one yellow seedling. (The sowings could be made on moist blotting paper). The pots are kept together in the dark until the seedlings are about 5 cm high when they are brought into the light. At this stage the two pots are indistinguishable; all the seedlings are of the same yellow colour. A second similar pair of pots should be sown two or three days after the first pair and kept in the dark until the demonstration is required, perhaps ten days after the first sowing. One of the first two pots now contains uniformly green seedlings after the exposure to light, whereas a quarter of those in the other pot are still yellow. The point is of course that even if a plant has the potentiality (the necessary genes) for producing chlorophyll, this potentiality can only be realised if the environment provides the right conditions (in this case light).

3 *Breeding investigations using barley*

The skill needed for the initial crossing and the two and a half year interval before the results from the F_2 generation are obtained are serious disadvantages in undertaking breeding investigations with barley, or indeed any plants whose life-cycle cannot be completed in less than one growing season. There is a method, however, by which results for analysis can be obtained from self-pollinated plants year after year without further crossing. The method is described here although it is equally applicable to peas (*Pisum sativum*) and any other self-pollinated species

Let us suppose that a dihybrid cross has been made between a pure black, awned variety and a white, awnless one. The F_1 will be all black and awned. When these seeds are sown in the following year the F_2 will give progeny in the ratio of 9 black, awned; 3 black, awnless: 3 white, awned: 1 white, awnless.

Suppose we now take one or two ears from each of say 32 of the plants showing the double dominant character and in the following year sow the seeds from these ears in separate rows, each row representing the progeny from a single plant. That is we shall have 32 rows with about 60 plants in a row. The space occupied will be no more than about 10 m x 5 m.

When ripe the rows will be found to be of four different kinds:

a) Those containing plants of all four phenotypes, that is, they were derived from a plant segregating at both loci and the individual plants would occur in the theoretical ratio of 9:3:3:1.
b) Those containing plants that were all black, but some were awned and some were awnless, that is segregation was at the awned/awnless locus only and the two phenotypes would occur in the ratio 3:1.
c) Those containing plants that were all awned but in the ratio of 3 black: 1 white and segregation has been at the black/white locus only.
d) Those containing only black awned plants like the double dominant original parent and no segregation has taken place.

The four kinds of rows would be in the theoretical ratio of 4:2:2:1. Plants which are black and awned, from rows of type (a) will provide the seed for the following year so that the demonstration is self-perpetuating. It also provides data for testing for the 9:3:3:1, the 3:1, and the 4:2:2:1 ratio.

4 *Artificial induction of mutations*

Barley is a convenient plant for mutation studies and it has been extensively used for this purpose.[15] The seed is an obvious stage at which to subject the plant to treatment. Any variety will do although in America much work has been done with the variety Himalaya, which has naked seeds. The seed can be treated dry or after soaking in water for periods up to 24 hours.

Gamma radiation in the kinds of dosages required cannot be used by schools but irradiated samples of seeds are available commercially. Suitable dose rates are 0, 5, 10, 15 and 20 rads.

Chemically mutations can be induced by ethyl-methane-sulphonate (EMS) in aqueous solution. Like all chemical mutagens EMS is carcinogenic and must be handled with extreme care. Rubber gloves must be worn when ethyl-methane-sulphonate is being handled and care taken not to inhale the fumes or allow drops to come in contact with the skin. All manipulations involving the concentrated liquid should be carried out in a fume cupboard and, while dilute solutions are probably quite harmless, it might be wise to leave the seeds in a fume cupboard during treatment.[9,12]

Immersion of the seeds in the selected dilution for the chosen time (6 to 24 hours) is carried out in muslin bags suspended by strings by which they are handled. After treatment and still in the bags the seeds should be washed in running water for upwards of two hours to remove all traces of EMS. After light drying on blotting paper the seeds should be planted out as soon as possible. Since only one gene from a pair of alleles is likely to have mutated, mutants will not as a rule be detectable in the first or M_1 generation. Records should however be made of percentage germination, percentage survival to three-leaf stage and percentage survival to fruiting. The seeds from each surviving plant are harvested separately and stored in labelled packets.

Test sowings of about 20 seeds from the packets can be made soon after the M_1 plants have been harvested. Any packets giving what might be mutants amongst the progeny from these test sowings are retained for further trials in the following year. Chlorophyll mutants are common and the easiest to recognise. Eleven different chlorophyll deficient forms have been described and some of them turn up regularly, especially the yellow Xantha and the white Albina forms.

The next step is to confirm that the suspected mutations are in fact inheritable and not the effect of treatment damage. To do this a larger test sowing from the remaining seeds in selected packets is made the following spring, the seeds from each packet making up an identifiable row. The surviving plants are grown to maturity and once again the seeds from each plant are harvested separately and stored in appropriately labelled packets. Test sowings will confirm whether or not the abnormalities were due to mutations and which plants were heterozygous. The seeds from the heterozygous plants are retained so that the mutations, almost all of which will be lethal, can be maintained in the heterozygous condition.

Wheat, Triticum spp. (2x = 14, 4x = 28 or 6x = 42)
1, 9, 20

Wheat, as one of the world's major food crops, has been intensively studied over a long period of time. Many of the advances in plant breeding have been developed in connection with wheat and some of the most sophisticated techniques of chromosome manipulation and cytological interpretation have been worked out on wheat materials. The evolution of wheat is of particular interest, not least because of its genetical implications. It has been reconstructed with a very high degree of authenticity by bringing together evidence from such diverse fields as archaeology, geography, botany, genetics, cytology and biochemistry.

Suggested demonstrations

Three demonstrations of general as well as genetical interest involving wheat can be staged in a school garden of very moderate size.

1 *The evolution of wheat*

Four features stand out:

a) The characters associated with the evolution of most cultivated grasses from their wild progenitors. In the wild state, easily dispersed fruits, or propagules, furnished with tough adherent coverings and coarse protective outgrowths and whose period of ripening is extended are an advantage. Under cultivation, fruits that can be readily harvested at one time and can be easily threshed are desirable. The evolution of wheat therefore has been away from a brittle rachis of the ear, caryopses with tough coats and rough awns and tillers ripening at different stages towards a tough rachis, free-threshing fruits and uniformly ripening ears.
b) The part played by natural hybridisation and polyploidy in the process. Cultivated wheats include a tetraploid and a hexaploid series which are both allopolyploid since their constituent sets of similar chromosomes or genomes, can be traced to three wild species, two of which are so-called goat grasses from the genus *Aegilops* and the third from the wild wheat *T. boeoticum.*
c) The third feature is that a huge number of varieties, or cultivars, adapted to local conditions over most of the temperate latitudes, or suited to different purposes, have arisen or have been consciously developed. One estimate puts the number at 1800. Wheats are grown in countries as climatically different as India, Australia, Mexico and Canada and can be used for making bread, biscuits, macaroni and whisky.

d) Finally, average yields, measured in metric tonnes per hectare, have been increased tenfold in the last two or three hundred years, and are still rising. Perhaps above all, the demonstration can be used to illustrate the work of the plant breeder in using genetical principles, together with a knowledge of agricultural practice, to raise yields by producing more prolific strains capable of responding to higher levels of fertiliser treatment and to reduce losses from pests and diseases by breeding resistant strains, while at the same time ensuring flour of a high quality.[87,92]

The evolution of wheat is shown diagrammatically in Fig. 6. A living demonstration can be set out similarly by growing five or six plants of each species in pots or small patches on open ground.

2 *Demonstration of continuous (quantitative) variation in wheat and its explanation in terms of the additive effects of three genes.*

Dwarf wheats are being much used in wheat breeding programmes in many parts of the world and spectacular results are being achieved in several developing countries.

Three major genes for dwarfness which have additive effects have been identified and the material is particularly suitable for demonstrating the multiple gene explanation of quantitative inheritance.

The material available consists of four strains, which in a recent trial made under English conditions had the characteristics shown in Table 11.

Although each variety has a very distinctive mean value for height there is considerable overlap in the distributions except in the case of the three gene

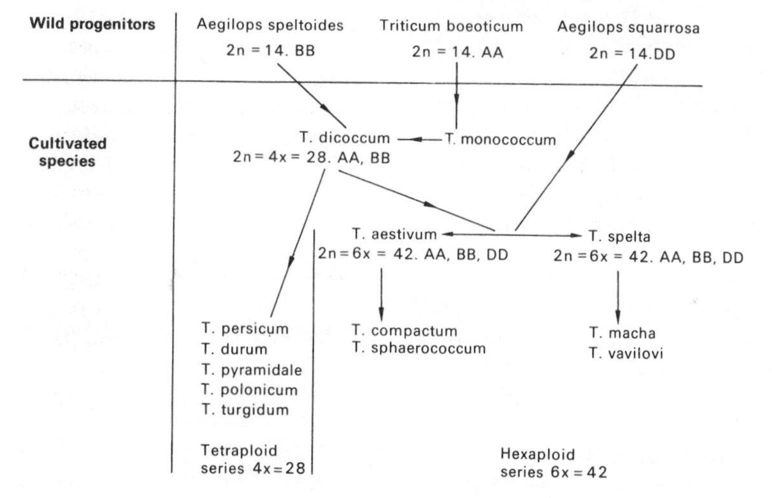

Figure 6 The evolution of wheat

Table 11 Characteristics shown by four strains of dwarf wheat under English conditions

Variety	Number of dwarfing genes	Number of plants	Height cm Mean	Range	Standard deviation
NP880 Normal	0	48	82	72–95	7.5
Lerma Rojo	1	50	76	64–84	5.0
Sonora 64	2	50	66	60–71	2.8
E5557	3	47	46.5	44–52	2.4

dwarf. The combination of the three genes has a very marked effect which may be due to the contribution of the third gene by itself or to interaction between genes. The presence of other genes affecting height cannot of course be ruled out.

It must be remembered that these strains have been selected for uniformity and in the trial they were grown under fairly uniform conditions. It is to be expected that if the four kinds of seeds are thoroughly mixed and sown unevenly on a piece of ground with a certain amount of variation in its physical conditions, the combined curve would be even smoother. A suitable demonstration, therefore, would be to grow say 50 seeds of each variety separately in uniform rows and a mixture of 50 seeds of each at random on a patch of uneven ground. Some criterion of the height to be measured for each plant must be adopted. In the trial reported above it was taken as the distance between the crown of the root system and the bottom of the tallest ear (primary tiller). These wheats are spring wheats and are best planted in Britain at about the middle of March. They tend to ripen at different times NP880 being more sensitive than the others to day length and taking considerably longer to reach maturity.

3 *In-breeding and out-breeding demonstrated by wheat (Triticum aestivum) and rye (Secale cereale)*

As a consequence of its outbreeding reproduction pattern samples of rye, even those grown from selected agricultural strains, show more variation than similar samples of wheat. It might also be noted that while there are some 1800 varieties of wheat currently listed, the corresponding figure for rye is only 80. These two facts are not unrelated.

Two demonstrations can be considered, one short term and simple to carry out, the other long term and involving some considerable effort but giving very convincing results.

Short term demonstration

Sow two fairly deep boxes, or better still, two small plots of ground with wheat and rye (commercial varieties) respectively. Record, quantitatively where possible, the appearance and growth of the seedlings and later, the maturing plants. The rye will be found to be much more variable than the wheat.

If the demonstration has been conducted in open ground some of the flowering spikes of each species should be enclosed in cellophane bags as soon as they begin to emerge in order to show that wheat is self-pollinated and rye cross-pollinated.

Long term demonstration

The idea here is to grow wheat and rye year after year, the seed used being some of that harvested the previous year. The piece of ground used should of course be changed from year to year perhaps as part of a rotation pattern being used for other purposes. To make the effects of the two breeding patterns more obvious small sowings of other wheat and rye varieties showing as many other distinctive characters as possible should be made surrounding the main plots so that inter-pollinations can occur where the breeding pattern permits.

In effect the different wheat varieties form separate populations and even under these favourable conditions variation hardly increases at all. On the other hand the rye varieties comprise one inter-breeding population and the variety being studied acquires increased variability.

The edible pea, Pisum sativum (2n = 14) *3, 4, 16, 18*

Suggested investigations and demonstrations

1 *Choice of the organism and characters for breeding investigations*

It was part of Mendel's genius that the organism he chose for his chief experiments, the edible pea, and the characters he selected for study lent themselves to his purpose.

The factors bearing on this question of the selection of suitable material can be listed as follows:

a) The organism should be a typical and representative one.
b) It should have forms that breed true with several pairs of contrasting characters.
c) These characters should be known to be inheritable and little affected by the environment.
d) The organism should have a short life cycle.
e) It should be easy to rear or to grow under fairly standardised conditions.
f) It should be easy to control mating and to bring about fertilisation.

g) It should be easy to protect the female parent from contaminating sperm or pollen.
h) Numerous offspring should result from a single mating.
i) Other things being equal, small organisms which are cheaper and easier to rear and take up less space are to be preferred to larger and costlier ones.

Mendel used seven pairs of contrasting characters. By a remarkable chance the genes responsible for these alleles are carried on the seven chromosome pairs, so that his work was not complicated by the phenomenon of linkage. These seven pairs of characters have been incorporated in two strains, one Pellew's gp (standing for green pod) exhibiting all the dominant characters except two in which it shows the recessive (seed shape and yellow pod) and the other, Cambridge Line 1, showing the alternative characters in each case. The characters used by Mendel and additional characters are shown in Table 12.

A very striking demonstration that provides much material for observation and discussion is to sow four large pots (about 25 cm) each with about six seeds of one of the following: Pellew's gp, Cambridge Line 1,

Table 12 The characters used by Mendel and the modern symbols used to represent them

	Mendel's description of the character	Dominant	Symbol	Recessive	Symbol
1	Difference in the form of the ripe seed	Round	RR	wrinkled	rr
2	Difference in the colour of the cotyledons	Yellow	II	green	ii
3	Difference in the colour of the testa	Coloured	AA	white	aa
4	Difference in the form of the ripe pods	Inflated	PePe	contricted	pepe
5	Difference in the colour of the unripe pods	Green	GpGp	yellow	gpgp
6	Difference in the position of the flowers	Axial	FaFa	terminal	fafa
7	Difference in the length of the stem	Long (Tall)	LeLe	Short (dwarf)	lele
	Additional characters shown in Pellew's gp and Cambridge Line 1				
1	Difference in the shape of the wings of the flower	Normal	KK	keeled	kk
2	Difference in the nature of the leaf	Normal	AcAc	acacia	acac
3	*Difference in the colour of the flower	Coloured		White	
4	*Difference in the form of the ripe seed	Drum/Round		drum/round	

*Several genes are reported affecting these two characters so that no symbols are given.

and the hybrids between them with Pellew's gp and Cambridge Line 1 as the female parent respectively.

2 *Monohybrid and dihybrid crosses*
Seed characters are the most suitable for this purpose since, as they are shown in the ripe seed, the results will be available at the end of the second growing season. The characters which Mendel emphasised and which are so often quoted in text books in reference to his work are the round/wrinkled and yellow/green seed characters listed as 1 and 2 in Table 12.

A most important point to remember when using seed characters for genetical investigations is that whereas the embryo and endosperm (for example, cotyledon colour and aleurone structure) are derived from fertilised egg cells and are thus part of the next generation, the testa is derived from maternal tissue and thus part of the parental generation so that a testa character will only be shown in the seeds of the following year's crop.

Once the original crosses have been made the experiment can be made self-perpetuating by the methods described for barley on pages 63 and 102.

A point that has to be borne in mind when planning breeding experiments with plants, and this is particularly true of peas, is to ensure that the parental plants are in flower at the same time so that crossing can be carried out. The choice of suitable varieties is one way of solving the problem. Another way is to make successional sowings, say at weekly intervals, of one of the parents.

The following cultivars have been found suitable:

Monohybrid cross: Sugar pea tall x Sugar pea dwarf.
Dihybrid cross: Alderman (green seeds, stringed pod) x Sugar pea (coloured seeds, stringless pod).
Little Marvel (yellow, wrinkled seeds) x Feltham Advance (green, round seeds).

3 *Three levels of expression of the genes responsible for the round/wrinkled seed characters*
The concept of the gene can be established from breeding investigations, particularly those involving total and partial linkage. This investigation suggests that the way genes act is biochemically.

This investigation is, of course, given greater point if it follows previous breeding experiments, either monohybrid or dihybrid, involving the round/wrinkled allelic pair of characters. In any case it needs to be stressed or, better still, supported by visual evidence, that round and wrinkled peas occur together in the same pod so that the difference is likely to be inherited and not environmental.

Round (smooth) seeds have elongated oval shaped starch grains typical of 'potato' starch whereas wrinkled seeds have more spherical, compound or 'split' grains somewhat subdivided. Both have as well a few small round grains. In addition there are differences in the chemical compositions of the two kinds of seeds. Round seeds have an average of 46.3 per cent starch, of which 38 per cent is amylose. Wrinkled seeds have an average of 33.7 per cent starch, of which 69 per cent is amylose.

The three levels of expression are:

a) the difference in the appearance of the starch grains,
b) the difference in the ability of the two kinds of pea to absorb water and
c) the difference in the activity of the starch forming enzymes in the seeds.

Suitable cultivars are Alaska (round yellow) and Little Marvel (wrinkled yellow).

Procedure

a) Examine, under a microscope, the starch grains after teasing a little of the cotyledons of the seeds in iodine solution.
b) Weigh a number of seeds and then soak in water. Dry the surface with blotting paper and reweigh. Calculate the average percentage increase in weight.
c) Prepare starch-free extracts from the two samples of peas.
    i) Grind 10g dried peas in 10 cm^3 of distilled water until no large particles remain.
    ii) Centrifuge to obtain clear solution free from starch grains.
    iii) Test a portion of the supernatant liquid with iodine to confirm that no starch is present.
    iv) Pipette starch free supernatant liquid into labelled container.
d) Prepare a petri dish with glucose-1-phosphate agar and carry out the test for starch.
    i) Mark a diametric line across the underside of the base so that the dish is 'divided' in two.

Make an identifying mark on either side of this line. Then add four drops of the appropriate supernatant liquid on the surface of the agar each side of the line.

ii) After 30 minutes and at 15 minute intervals after that, test the area beneath the drops of supernatant liquid for starch by adding a drop of iodine solution. As a precaution in case some starch is still present in the extract remove the drop from the surface of the agar with filter paper and look for the presence of starch within the agar. The extract from the wrinkled peas is more active than that from the round peas.[15,17]

4 *Extraction and examination of deoxyribonucleic acid (DNA) and ribonucleic acid (RNA) from root tips*

The origin of commercial sources of DNA and RNA is commonly yeast or bovine thymus, spleen or sperm. It is possible, however, to extract them from root tips which has the advantage that besides being cheap and simple to work with the products can be related, by their staining properties, to the chromosomes and other cell inclusions revealed in squash preparations of dividing cells.[15,17]

A sequence of three investigations concerning DNA appears crucial to the establishment of modern genetical theory on an empirical basis at secondary school level.

The three steps in the sequence are:

a) the identification of DNA and RNA in cell preparations,
b) the extraction of DNA and RNA from pea root tips and
c) the hydrolysis of DNA and the identification of the nitrogenous bases.

a) *The identification of* DNA *and* RNA *in cell preparations*

DNA and RNA show up better in plants that have fewer and larger chromosomes than the edible pea. For this reason the preparations should be made with root tips of the broad bean (*Vicia faba*) or some other suitable alternative as well as the pea.

The Feulgen method: The Feulgen stain (leuco-basic fuchsin) is commonly used in chromosome preparations and is a specific biochemical test for DNA. It is based on Schiff's reagent, used in organic chemistry as a test for aldehyde (-CHO). DNA liberates aldehyde when it is hydrolysed.[7]

i) Fix in acetic alcohol for a half to one hour
ii) Rinse with two or three changes of water
iii) Macerate by hydrolysis in N HC1 solution at 60°C for about six minutes, or in a 50% aqueous solution of the concentrated acid at room temperature
iv) Stain in Feulgen stain for one to three hours
v) Cut off the tip of the root and tease out. (A drop of acetic-carmine stain may be added to intensify staining if necessary).
vi) Apply a coverslip and squash gently under several layers of blotting or filter paper
vii) Examine under a microscope.

Details of how the preparation may be made permanent are given on page 48.

The methyl-green-pyronin method: This mixed stain stains DNA green and RNA red. It is very convenient for the purpose at issue but it is not biochemical in its action or specific to either DNA or RNA. This is the reason for using the Feulgen stain as a check.[7]

i) Fix in absolute alcohol
ii) Cut thin sections from the root tip and stain in methyl-green-pyronin for about half an hour. (Teased tissue could be similarly stained).
iii) Wash and mount in water
iv) Examine under a microscope

If the whole roots are fixed and stained a day or two before they are required, valuable time is saved and the appearance of the roots, with only the tips stained, is a vivid indication of the concentration of DNA in the root tips which is the basis of the following investigation.

b) *The extraction of* DNA *and* RNA

The separation of the DNA and RNA from the other cell contents and from each other is effected by a combination of maceration, filtration (through very fine mesh bolting nylon or specially fine glass filters) and by centrifugation. In order to prevent hydrolysis all the solutions and utensils must be thoroughly chilled in a refrigerator before use and the operations carried out as far as possible in vessels surrounded by packed melting ice. The solutions must also be maintained at the correct pH value by means of buffers. These two points are useful reminders of the characteristics of enzymes.

i) Carefully wash the roots of about 50 germinating peas with roots about 3 cm long.
ii) Using a sharp razor blade and a piece of graph paper as a guide, cut the terminal 0.3 cm from each root.
iii) Place the tips immediately into a test tube containing about 10 cm^3 of cold buffered formaldehyde, pH7.3. Leave the tube in ice for 45 minutes.
iv) At the end of this time remove the tips, with a very small volume of the buffered formaldehyde, to a cold mortar and grind them thoroughly in the cold.
v) Filter the ground up mixture through a small piece of bolting nylon or sintered glass crucible straight into a cold centrifuge tube using slight pressure, or a vacuum if one is available.
vi) Centrifuge the filtrate at low speed (about 500 g) for five minutes.
vii) Withdraw a drop of the suspension from the bottom of the tube by means of a fine pipette and place it on a microscope slide.
viii) Add a drop of methyl-green-pyronin stain; leave for 15 minutes and examine under high power.
ix) Repeat, but this time remove a drop of the suspension from the top of the tube.

The suspension from the bottom of the tube contains a high proportion of nuclei stained green and hence a rough sample of DNA; the suspension from the top contains ribosomes stained red, amongst other small unstained fragments, and indicates a crude sample of RNA.

Having now given an indication of how DNA and RNA can be obtained the next step is to show something of the nature of DNA especially with reference to genetical coding, using purer specimens obtained commercially.

c) *The hydrolysis of* DNA *and the identification of the nitrogenous bases*

After hydrolysis of DNA, the hydrolysate is separated into its components by paper chromatography. The nitrogenous bases are then identified amongst these components by comparison against the chromatograms of the bases obtained from commercial sources.

There are thus three steps: hydrolysis, separation and identification of the component bases.

Hydrolysis

Perchloric acid must be treated with extreme care and must not come into contact with any organic material such as hair, clothing or cotton wool.

i) Weigh out about 5 mg DNA and add to 0.4 cm^3 distilled water and 0.1 cm^3 72 per cent perchloric acid in a small test tube. Use small teat pipettes for transferring the perchloric acid.
ii) Keep the test tube and its contents in a water bath at 100°C for 1 hour
iii) Cool it on ice and adjust the pH to about 3.5 by means of potassium hydroxide solution and pH paper
iv) Centrifuge to precipitate the perchlorates

The nitrogenous bases are left in solution. Further information may be found in reference 17.

Separation

A 'one-way separation' is needed for the identification of the bases, a 'two-way separation' effects complete separation.

One-way separation

i) Mark the position of the spots on the chromatographic paper about 2.5 cm from one edge. A suggested sequence of distribution on 20 or 25 cm paper is: hydrolysed DNA; guanine; adenine; DNA solution; cytosine; thiamine; hydrolysed DNA.
ii) As at least 5μg of each substance is needed for a 'spot' of it to show up in ultra-violet light, at least twenty applications of the solution are needed. These are best made with a small platinum wire loop as the 'spot' should not be more than 0.5 cm in diameter.

Care must be taken to allow the 'spot' to dry between applications and for the chromatogram to dry without heat after the run is completed. An electric fan can be used.

iii) Allow isopropanol-hydrochloric acid solvent to rise to the top of the paper, this may take up to eight hours.

iv) Dry the paper thoroughly.

Two-way separation

i) Proceed as for the one-way separation steps (i) to (iv) inclusive.

ii) Rotate the paper through 90° and allow the second solvent, butanol-ammonia, to rise to the top of the paper.

iii) Dry the paper thoroughly.

Occasionally a fifth 'spot' shows up. This can be shown to be uracil and the explanation is that it is not a constituent of DNA but the product of a further hydrolysis of the cytosine.

Identification

The chromatogram 'spots' are shown up by their fluorescence in ultra-violet light so that a suitable ultra-violet lamp and access to a dark room are required. A lamp with maximum emission at 260μM is best.[9,12]

A teaching point to be borne in mind is the connection between the fluorescence of the nitrogenous bases, DNA and RNA in ultra-violet light and the mutagenic effects of this radiation (see page 83).

Each fluorescent 'spot' is ringed with pencil. The distance each chemical has moved from the point of origin, relative to the distance the solvent has moved, is then measured and compared. The symbol used to designate this relative distance of travel is Rf.[25]

The Rf values of the bases, when isopropanol-hydrochloric acid is used as a solvent at 15°C, are:

Guanine	Rf = 0.22 or 22%
Adenine	Rf = 0.32 or 32%
Cytosine	Rf = 0.44 or 44%
(Uracil	Rf = 0.66 or 66%)
Thymine	Rf = 0.76 or 76%

Broad bean, Vicia faba (2n = 14) *7, 13*

The broad bean has a very long history as a food plant dating back to at least 7000 BC. It is now unknown as a wild plant. It has large chromosomes and is particularly good for cytological preparations of root tips.

Suggested demonstrations

1 To obtain good tips for squash preparations soaked seeds should be grown suspended over water or, perhaps better still, in vermiculite. If the tip of the primary root is cut off when the root is about 3 cm in length a good crop of thin lateral roots will develop in a few days. A few beans will therefore provide suitable material for a whole class.

2 Another use for the broad bean, requiring the pollinated flowers in this case, is to show the giant chromosomes that are to be found in the suspensor of the embryos within the fertilised ovules. These are comparable to the polytene chromosomes in the salivary glands of many dipterous flies such as *Chironomus, Drosophila* and *Simulium*. For details of the relevant staining techniques see page 45.

Sweet pea, Lathyrus odoratus (2n = 14) *4, 16, 20*

The sweet pea is worth consideration because its evolution, since it was first introduced into Britain in 1699, is so well documented and because some of the early work which extended the principles Mendel established was done with it. For example, gene interaction was exemplified as long ago as 1905 when two white-flowered plants of the variety Emily Henderson, were crossed giving a uniform F_1 with coloured flowers. This suggested that two factors, or genes as we should now say, acting together were responsible for the production of colour. Even earlier, in 1901, a famous example of a mutation occurred on three separate occasions. In each case it was the in variety Prima Donna and gave the large, waved upright standard which characterises the modern 'Spencer type' sweet pea in place of the smaller 'hooded' standard of the earlier 'Grandiflora type'.

Suggested investigations
Another mutation was to the dwarf or 'cupid' form and this suggests two genetical investigations.

1 Monohybrid cross: tall x dwarf.
2 Effect of growth hormones on dwarf plants. Treat dwarf plants with gibberellic acid and/or indoleacetic acid to try to produce the tall phenotype and to investigate the inheritance of the induced phenotypic tall character (see investigation with *Zea mays* page 60).

The inheritance of flower colour, flower form, length of the peduncle and the number of flowers on the peduncle could all be investigated. The last three are of course important show characters and much seeet pea breeding has been done in this connection as well as with flower colour.[9 1]

Groundsel and Ragwort, Senecio spp. (2n = 40) *4, 9, 11*

Common groundsel, S. vulgaris
This has a mutant form with a few (about 8) ray florets (var. radiatus Koch) which is gradually becoming more abundant and spreading across Britain from west to east). The F_1 between the wild type and this mutant is intermediate between the two and has a few short rays (about 3 mm) and is therefore easily recognisable. Fruits collected from these F_1 plants segregate 1:2:1 so that a supply of the F_1 material is assured.

In a garden where both forms occur, and this can be contrived artificially, it is interesting to record the relative abundance of the three forms at different times of the year and from year to year. Cross pollination is rare but once an intermediate form appears the population appears to move towards a balanced polymorphism and opportunities arise for studies in population genetics.

Senecio cambriensis, a natural allotetraploid
This appears to be a natural hybrid between *S. vulgaris* and Oxford ragwort (*S. squalidus*) with a doubling of the chromosome complement. It is significant that *S. squalidus* was at one time known in Britain only on the walls of certain Oxford colleges, having probably escaped from the Oxford Botanic Garden. Since the war, possibly because of the abundance of bombed sites, which provided a suitable habitat, it has spread rapidly, so that opportunities for hybridisation have increased.

Biennials

Crop plants illustrating important genetical principles in their evolution
Two simple garden demonstrations are given as illustrations.

Beets, Beta spp. (2n = 18) 20
The cultivated beets which include those grown for their leaves (spinach beets and fodder beets) and for their swollen roots (garden beets and sugar beets) are all derived from the sea-beet (*B. maritima*) a common plant of the seashores of Western Europe. They have been cultivated since at least Roman times. They are classified as *B. vulgaris*. The main theme is, of course, selection on the basis of multiple genes affecting characters such as leafiness, root shape and size, and sugar content but recently 'monogerm' varieties, that is those with only one fruit in the glomerules, and polyploid varieties have been put on the market. A suitable lay-out might be as given in fig. 7.

Brassicas: cabbages, swedes and turnips (2n = 18, 2n = 38, 2n = 20) 9, 20
These plants can be used very effectively to illustrate the wide variation to be found within a species, and also polyploidy. The swedes (*B. napus*) with a chromosome number 2n = 38 obviously resulted from the hybridisation of a cabbage (*B. oleracea*) 2n = 18, with a turnip (*B. rapa*) 2n = 20. Within each species there is a wide variation of cultivated forms resulting from selection by man over a long period of time but this affects primarily the vegetative features and the flowers, fruits and seeds are remarkably similar and the varieties inter-fertile. A suitable lay-out for a garden demonstration is shown in fig. 8 but many other alternatives could be found.

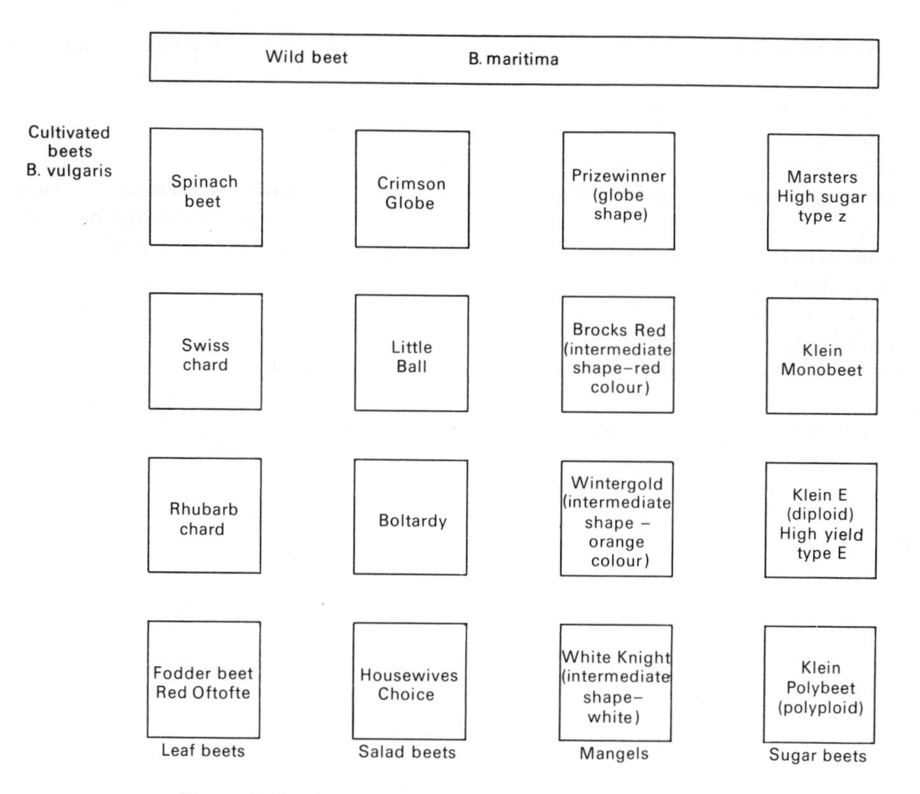

Figure 7 A suitable garden layout for a display of beets

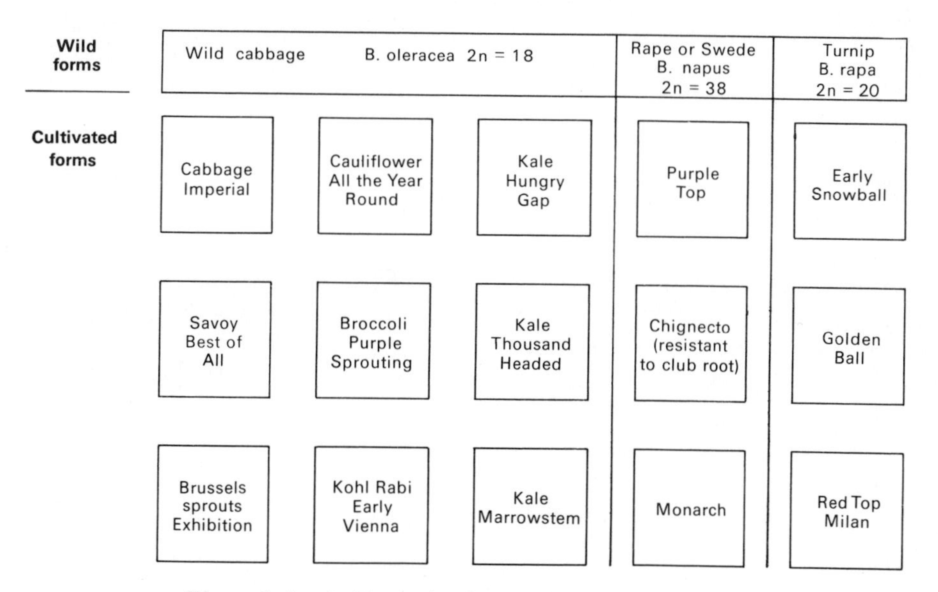

Figure 8 A suitable garden layout for a display of brassicas

Crop plant useful for several purposes

Radish, Raphanus sativa (2n = 18, 36). 1, 4
The familiar salad radish has been cultivated for so long that its origin is lost in antiquity. It is not known as a wild plant. Genetically it can be used for two purposes.

1 To show variation in colour and shape of roots e.g. Cherry Belle, small, round red roots; French Breakfast, large, long red roots with white tips; Icicle medium, long and tapering, white; Mino Early, large, long red, rapidly maturing; Black Spanish, long, black and strongly flavoured.
2 Dihybrid cross with both characters, colour and shape, giving rise to intermediate forms. A suggested cross, both ways is Scarlet Globe (red, round) x Icicle (white, long).

Biennial garden plant for breeding experiments giving seedling characters

Wallflower, Cheiranthus cheiri (2n = 14) 4
Two varieties, the yellow Cloth of Gold and the deep crimson Vulcan differ by a single gene, which, in the homozygous recessive inhibits the formation of anthocyanin. When the two parents are crossed all the seedlings in the F_1 have red hypocotyls and as mature plants bear red flowers. When these red heterozygotes are selfed, the seedlings in the resulting F_2 segregate 3 red hypocotyls: 1 white hypocotyls. The pigments can be extracted and separated by paper chromatography (see page 70).

8 Flowering plants and ferns – hardy field and garden plants, perennials

Hardy perennial plants of genetical interest fall into two major groups so far as schools are concerned, those that illustrate a genetical phenomenon but which at the same time serve as decorative features of the school grounds and those that are required from time to time to provide materials for a specific purpose. Plants showing somatic mutations or chimaeras are examples of the first group; plants providing suitable material for chromosome preparations are examples of the second. In either case the needs of the plants with respect to space, soil, moisture and light as well as their place and appearance in the garden, have to be taken into consideration when they are planted. A brief outline of these needs is given in Table 17, page 101. Full details of the requirements of the plants can be obtained from references 90, and 97 or the sources from which the plants are obtained. When a school garden is being laid out the possibility of providing teaching material as well as for making a pleasant setting for the school should be borne in mind. The range of possibilities is very great and clearly there will be other needs to meet besides those of biology.

Decorative Plants of Genetical Interest

Roses Rosa spp. (a classical polyploid series in which x = 7. 2x, 3x, 4x, 5x and 6x species are all known) *10*

Roses require comparatively little attention once planted and can find a place in most school gardens. For further information see references 95 and 102, specialist advice may be sought from the Royal National Rose Society.

1 **Possible line of evolution of modern roses**

The course of the evolution of the various kinds of modern roses is not known with any certainty and it is now impossible to obtain some of the wild species since only their cultivated derivatives have survived. Nevertheless it is instructive to grow together some of the species which are believed to have contributed to, for example, the modern hybrid teas. These are:

R. alba (The White Rose of York, though not a strict species is also worth a place in this collection. Height about 2m).

R. centifolia var muscosa 4x (Common Moss rose. Height 1.5 to 2m).

R. chinensis 2x (The common China rose—the species has been lost but a primitive form, Old Blush is still obtainable. Height 1 to 1.5m).

R. damascena 4x (The Damask rose—fragrant and the source of attar of roses. Height 1 to 1.5m).

Rosa gallica var officinalis 4x (The Red Rose of Lancaster. Height 1 to 1.5m).

2 **Origin of new cultivars**

The way in which new cultivars are produced by mutation, hybridisation (natural and deliberate) and selection can be demonstrated:

2.1 *Mutation*

Ophelia was the starting point of a number (all about 1 m high):

Ophelia (pale pink, 40 petals, very fragrant, few thorns, vigorous) by mutation gave,

Madame Butterfly (pale pink with yellow shading, 35 petals, very fragrant, long stems, few thorns, free flowering, vigorous) which by mutation gave, Lady Sylvia (pink with yellow shading, 33 petals, very fragrant, few thorns, exceptionally free flowering, very vigorous).

2.2 *Hybridisation—natural*

Golden Ophelia (pale yellow, vigorous), a natural seedling from Ophelia

2.3 *Hybridisation—deliberate*

Ophelia x Souvenir de Claudins Pernet gave Talisman (scarlet with pink and gold shading, 30 petals, fragrant, free flowering, moderately vigorous).

3 **Effect of stock on scion (environmental variation)**

The same buds or grafts grown on different stocks will show considerable variation.

4 Gene mutation—unstable genes

Rosa gallica var. *versicolor (Rosa mundi)* (Height 1 to 1.5m) A very early mutant from *Rosa gallica* with large semi-double blooms of light crimson heavily splashed and striped with pink and white, the consequence of somatic mutations.

5 Genetical reversion

Climbing roses often revert to bush roses and vice-versa. These are well worth keeping to demonstrate reversion.

Shrubs and Trees: 'Normal' wild type and mutant forms *10*

Hazel, Corylus avellana and *willows, Salix spp.* and their *distorta* varieties with tortuose twigs and leaves. (Height 2m and over).

Juniper, Juniperus communis (Height 1 to 6m) has two geographical forms, the lowland form s.sp.*communis* which is prickly to touch, erect habit and a globose fruit and the mountain form s.sp.*nana* which is scarcely prickly, procumbent habit and an elongated fruit.

Dyers greenweed, Genista tinctoria (Height 1m) which is common throughout England and Wales has a procumbent variety var. *humifusa* found only near the Lizard in Cornwall and St David's Head, Pembrokeshire.

Jews mallow, Kerria japonica (Height 2 to 3m) has a double form var. *flore plena* and a variegated dwarf from var. *variegata.*

Common ling, Calluna vulgaris (Height 0.5m) has a grey tomentose wild form very common in some parts of Scotland var. *hirsuta* and easily obtainable garden varieties var. *searlei* (white flowers), var. *aurea* (golden leaves), var. *argentea* (silvery leaves) and two with double flowers. H.E. Beale (tall) and J.H. Hamilton (dwarf). All these need a lime-free soil.

Mountain ash, Sorbus aucuparia (Height to 10m) has a yellow fruited form var. *xanthocarpa*, another with deeply cut leaves var. *asplenifolia* and another more upright in growth var. *fastigiata.*

Natural and Garden Hybrids *9*

The parents and their hybrid, and if the latter is fertile also the backcrosses can be grown together so as to bring out their relationship. Occasionally some other features such as a mutant form can be added. By convention the symbol x before a generic or specific epithet signifies that the plant is of hybrid origin.

Natural hybrids

Primula spp. (Height to 0.5m) The occurrence of heterostyly in the primrose (*P. vulgaris*) is well known but the genetical basis is not so well known and this, and also the distribution of heterostyly in the various sections of the genus, offers scope for study at school level.

A common hybrid, the false oxlip between the primrose (*P. vulgaris*) and the cowslip (*P. veris*) can usually be found where primroses and cowslips grow close together. The true oxlip (*P. elatior*) which is fairly rare plant now found only in a few woods in East Anglia also hybridises easily with the primrose and numerous hybrids and backcrosses are usually to be found where they grow together. All have the same chromosome number 2n =22 and they set seed freely so that they can be grown from seed.

The bird's eye primrose (*P. farinosa*) which grows on wet basic soil in the north of England and Southern Scotland belongs to a group of this genus which consists of a polyploid series. The members of this series make an interesting garden demonstration. Thus:

x = 9	2x = 18.	*P. farinosa, P. frondosa, P. modesta*
	4x = 36	*P. longiflora, P. yupaensis*
	6x = 54	*P. scotica.*

Red campion, Melandrium rubrum (Height about 0.5m) and white campion *M. album* (Height about 0.5m) cross to give fertile hybrids and backcrosses which, as the colour of the offspring is intermediate between the parents, can be grown to show a graded series. As the flowers of both species are dioecious there is obvious scope for breeding experiments. (2n = 24 in both species)

Wood avens, Geum urbanum and water avens G. rivale give a hybrid *G. x intermedium.* There is also a white mutant of *G. rivale* (2n = 42) in all cases; height about 0.5m).

Docks. Although they are hardly garden plants a space might be found for two docks and the very common almost sterile hybrid between them. They are the *curled dock, Rumex crispus* (2n = 60) and the *broad leaved dock R. obtusifolius* (2n = 40) Height about 0.5m.

Rushes, Juncus spp., are neglected in school courses but two pairs which are fairly easily recognised give common completely sterile hybrids which are intermediate in many characters between the two parents. They are *J. inflexus* (hard rush, 2n = 40) x *J. effusus* (soft rush, 2n = 40) and *J. articulatus* (jointed rush, 2n = 80) x *J. acutiflorus* (sharp flowered rush. 2n = 40). Height 0.5 to lm.

Garden hybrids

Fatshedera x *Lizea*, a slightly tender evergreen shrub, height about 2m. This is an intergeneric hybrid obtained by crossing *Hedera hibernica* (Irish Ivy) with *Fatsia japonica* (Figleaf Palm).

x *Gaulnettya wisleyensis* is a hardy evergreen shrub which grows to a height, of about 0.6m. This is an intergeneric hybrid between *Pernettya mucronata* (Prickly Heath) and a species of *Gaultheria.*

Foxgloves, Digitalis purpurea (Height about 1 m) are biennials and in an attempt to produce a perennial garden form they were crossed with a yellow perennial mountain member of the genus quite common in the European Alps. This is *D. grandiflorum* and the strawberry pink coloured perennial hybrid is *x D. mertonensis.* Together with these three plants, all easily grown from seed, the fairly common white mutant form of *D. purpurea* and some of the very handsome garden forms obtained by selection can be grown to give an interesting feature in the garden. Another mutant, with the terminal flower of the spike of a large open or peloric form, is also fairly easy to obtain and grow from seed.

An attempt to improve the wild blackberry, Rubus spp. (x = 7). Two wild species *R. thrysiger* (4x = 28) and *R. rusticanus inermis* were crossed to give a fertile hybrid known as the John Innes Berry (4x = 28) which was later improved by selection to give Merton Thornless. *R. thyrsiger* has a long, well-spaced inflorescence and *R. rusticanus inermis* is a mutant form of one of the commonest blackberries in the south of England, which is thornless. The best features of the two parents have been brought together in the 'hybrid'.

Plants illustrating polyploidy *9*

The primroses, roses and wheat have already been described (p. 54, 75 and 64).

Polyploidy associated with geographical distribution

Valerian, Valeriana officinalis. (Height about 0.5 to 1m). The tetraploid 4x = 28 is found on chalk and limestone in the southern and midland countries of England and the octoploid 8x = 56 in damp valleys in the south and in both dry and damp habitats in the north.

Polyploidy associated with pollen size

Bladder campion, Silene cucubalus. (Height about 0.5m, 2x = 24 and 4x = 48) Though natural hybrids are rare *S. cucubalus* (2n = 24) forms fertile hybrids with *S. maritima.*

Polyploidy associated with ecological conditions and fertility

Ranunculus ficaria var. fertilis (2x = 16), leaves with no axillary bulbils, petals broad and overlapping, largely fertile—the commoner form and found in sunny places. Var. *ficaria* (4x = 32), leaves with axillary bulbils that reproduce the plant vegetatively, rarely fertile—found in damp, shady places.

Polyploidy in ferns associated with geographical distribution and differences in fertility
Polypodium australe (2x = 74), very local on limestone in Britain
P. vulgare (4x = 148)
P. interjectum (6x = 222), more robust than the other two. Heights about 1m.

Naturally occurring triploid and related species from North America
The modern garden forms of *Tradescantia virginiana* are triploids and tetraploids (3x = 18 and 4x = 24). They are very different from the three wild diploid species *T. ohioensis*, *T. virginiana*, and *T. pilosa* from which they are now thought to have been derived. Other diploid species are *T. paludosa*, *T. brevicaulis*, and *T. brachteata* (2n = 12). Heights between 0.5 and 1m. The flower buds of all these species are useful for meiosis squash preparations. For procedures see pages 45 & 47.

Plants illustrating chimaeras or various forms of variegation
Chimaeras can often be seen in variegated plants, branches of which appear fully green or white. *Laburnocytisus* is a graft hybrid of broom (*Sarothamnus scoparius*) over laburnum (*Laburnum anagyroides*). Branches of the hybrid sometimes produce branches similar to one of the parental species. Height up to 5m.

Plants maintained in the garden for occasional use *7, 8*
So far as genetics is concerned the main need is for plant material for chromosome preparations, that is for root tips and pollen mother cells from species which have large chromosomes, preferably with distinctive characters, and a low chromosome number. Success in making good chromosome preparations depends upon fixing suitable tissues when they are actively dividing. A point to bear in mind too is that at some point students should study mitosis and meiosis in the same organism so as to be able to check the relationship between the haploid and diploid numbers.

Other plants that regularly show chromosome abnormalities are also worth growing, as are diploids and tetraploids of the same species.

The perennial plants for this purpose may be conveniently divided into the bulbous and the non-bulbous species. Details of the squash method of chromosome preparation are on pages 45 & 47.

Bulbous species
Root tips are best obtained by transferring the bulbs or corms just before their growing season begins to pots of vermiculite or to stand on wet gravel or to be suspended just above the water level in a jar of water. They should be kept in the dark. Flower buds are best taken from actively growing plants in their natural surroundings and it is a good plan to take and fix them at three or four different times of the day and night, as many plants show a daily rhythm in meiosis. Choosing buds of the right size is a matter of trial and experience so that, other things being equal, plants bearing many flower buds at different stages of development are to be preferred. The anthers within a bud are usually synchronised so that once a suitable anther has been recognised the remainder from the bud can be put aside for further use.

Allium spp. (2n = 16) *7*
Some of the decorative species. e.g. *A. moly*, are good but most people use the onion *A. cepa* or garlic *A. sativum*, in which case the bulbs are best bought from greengrocers' shops for the purpose if root tip preparations are in mind. There is a difficulty in getting roots to grow when the bulb is dormant in the autumn. This can be overcome by keeping the bulbs in a refrigerator at 4°C for several weeks before using them. (Height less than 0.5m).

Crocus balansae var. Zwanenberg (2n = 6) *7*
For root tips, the corms will produce vigorous roots if dug up in September and grown indoors in vermiculite. They will be in good condition November to January. The three pairs of chromosomes are easily distinguishable.

Bluebell, Endymion non-scripta (2n = 16) *7, 8*
Can be used for root tips and pollen mother cells in January/February. The flower buds have to be taken from inside the bulb when the shoot is about 5cm tall.

Hyacinthus spp. (2n = 16) *7, 8*
As for *Endymion* but one to two months earlier.

Non-bulbous species

Red-hot pokers, Kniphofia spp. (2n = 10. Height about 1m.) *8*
Good root tips are not easy to obtain but buds for pollen mother cells are available in June.

Peonies, Paeonia spp. (2n = 10. Height 0.5 to 1m)*8*
Here again the plants are not suitable for root tips but the single species are all diploids and are very good for pollen mother cells in April and May. Buds of about 1.5cm are the right size; they fix well and can be kept in 70% alcohol for several months. Each flower bud contains numerous stamens and the buds can be collected at any time of the day as meiosis is not rhythmical. Chromosome abnormalities can often be found.

Tradescantia spp. (2x = 12. Height 0.5 to 1m) *7, 8*
Tradescantia has been referred to in the previous section in connection with polyploidy. The diploid species make excellent material for both root tip and pollen mother cell preparations. To obtain good root tips small portions of a vigorously growing plant should be potted up in compost in small pots and kept well watered for several weeks. The root tips can be cut off from the outside of the soil ball when the plant is 'knocked-out' of its pot. Alternatively, stem cuttings can be rooted in jars of tap water. Light should be excluded from the part of the stem in water. Buds for pollen mother cells should be about 0.3–0.5 cm. in length.[7,15,99]

Clones of white clover, Trifolium repens (4x = 32) *1, 2, 15, 16*
T. repens is a wild tetraploid whose ancestor or ancestors are not known. It is an important forage legume indigenous to Western Europe from which selected strains have been introduced into many parts of the world. Clonal material can be maintained as the plant reproduces vegetatively by creeping stolons, and hybridisation is simple as the species is almost completely self-sterile.

Three demonstrations can be made: (1) the effect of environmental factors and interspecific competition, (2) genetical variation, (3) cyanogenesis and selective predation.

Different phenotypes can be collected from wild populations in permanent pastures and sports grounds or from agricultural varieties in leys, or they may be sown from seed of named varieties. The best method of growing them is in large pots of suitable compost sunk into the soil.

Environmental variation
Any environmental factors can be varied, soil, moisture, light intensity, although competition in a natural sward suggests that the response of the plants to different light intensities might be the most instructive. Competition experiments between large-leaved and small-leaved varieties give interesting results.

Genetical variation
Besides the large and small leaved varieties which give a continuous range from one extreme to the other there are two major multiple alleles that give a clearly discontinuous series in both cases.

1 white markings due to shape and distribution of the underlying leaf cells—the V series.
2 red markings due to the presence of anthocyanin—the R series.

Cyanogenesis
The study of cyanogenesis makes two valuable contributions to the systematic building up of genetical theory in schools, firstly it gives a clear example of one particular gene being intimately related to, one could almost say being responsible for, the production of a specific enzyme, and secondly it demonstrates the interaction of two genes to produce an effect that can be shown to have a selective function.

Briefly, cyanogenesis involves the release of hydrogen cyanide from a mixture of cyanogenic glucosides, either rapidly through the action of an enzyme, linamarase, or slowly by hydrolysis. In *T. repens*, and many other plants, four phenotypes, corresponding to the four genotypes obtainable from two genes displaying dominance, can be recognised. The genes are Ac ac, glucoside present or absent, Li li, enzyme present or absent.

The four phenotypes are as follows:

1 Plant leaves release HCN rapidly when maccrated in a drop of water and incubated at 40°C (Ac-, Li-).
2 Leaves release HCN slowly, in about 24 hours, when similarly treated, or more quickly if the enzyme extracted from leaves known to contain it is added (Ac-, lili).
3 Leaves do not release HCN unless the glucosides, which can be extracted from leaves known to contain them are added (ac ac, Li-).
4 Leaves do not release HCN even when the glucosides are added (ac ac, lili).

The advantages of maintaining the four phenotypes as clonal material are that students can practise the techniques with them before undertaking population studies in wild populations and that the extracts of the enzyme and the glucosides can be readily prepared from them. See references 6, 14, 16 for further information.

9 Micro-organisms– fungi and bacteria

Since fungi were first used for genetical studies in about 1930, and more especially since Beadle and Tatum published their work on biochemical mutants in 1945, a vast amount of work has been done with them and many of the most significant recent advances in genetics have been made through them. Their usefulness rests on a number of characteristics including the nature and shortness of their reproductive cycles, the ease with which their nutritional requirements can be studied and controlled, the readiness by which they can be propagated from small pieces of mycelium or spores and the convenience with which they can be grown under standardised conditions.

The species used are almost invariably saprophytic in their nutrition. They are normally cultured in a nutrient liquid or on a nutrient gel plate, usually an agar plate. As a rule the nutrients are completely defined and as fungi are unable to use carbon dioxide as a source of carbon, the nutrients must include a carbon source, such as sugar, together with a nitrogen source, often in the form of an amino acid, and vitamins and inorganic salts.

Many thousands of fungal mutants have been found or artificially induced. These may be roughly classified as morphological, nutritional or biochemical and respiratory mutants, although there are some others.

In the case of nutritional mutants, which for our purposes are probably of the greatest use, the culture medium employed can take one of several forms.

Minimal medium is the simplest one on which the wild type, or prototroph, fungus can grow. It often contains glucose (as a carbon source), an amino acid, a vitamin and inorganic salts.

Complete medium is one that has been enriched with organic products such as yeast extract, malt extract, casein and nucleic acids and on which the nutritional mutants will grow vigorously. Normally both wild type and mutant stocks are maintained on complete medium. Nutritional mutants requiring some auxiliary substance above those found in minimal medium for satisfactory growth are called auxotrophs.

Other media are made by adding controlled proportions of various nutritional supplements, often amino acids or vitamins, to the minimal medium.

At school level fungal genetics is probably best used to illustrate the phenomenon of complementation and for tracing biochemical pathways, that is, for the study of gene action. Ascomycetes with a linear row of spores in their asci such as *Sordaria fimicola* and *Neurospora crassa* are, in addition, valuable for demonstrating the mechanism of crossing over and the principles of chromosome mapping. Suggestions for further work using fungi and bacteria will be found in references 107, 108, 109 and 110.

Management

See pages 104 & 105.

Fungi

Yeasts *10, 16*

Yeasts are a somewhat anomalous group of organisms but there is little difficulty in recognising the commoner ones for what they are, namely unicellular fungi. In addition to vegetative reproduction by budding many yeasts undergo a sexual phase in which asci containing four ascospores are formed showing the affinities of these yeasts with the Ascomycetes. In some species the ordinary vegetative cells are haploid (n) and two vegetative cells conjugate (2n) before ascus formation. In such cases the four ascospores, after release from the ascus, germinate and bud as vegetative cells. In other species the vegetative cells are diploid (2n) and meiosis occurs in the formation of the ascospores within the ascus. In these cases conjugation between ascospores usually occurs soon after the spores escape from the ascus, although haploid strains can be cultured and form the basis of much of the work.

The species most commonly used in school work, and the one we shall describe here is *Saccharomyces cerevisiae;* it belongs to the second group. The spores are of two mating types, a and α, two of each type being formed in each ascus. Conjugation can only occur between spores or cells of different mating types. It will be obvious that this fact can be put to good use in genetical experiments by allowing crosses to be made quite simply between haploid colonies of two different mutants or strains so long as they are of different mating types. The life cycle of *S. cerevisiae* is shown in fig. 9.

Besides its uses in demonstrating fungal life cycles with mating types and for work on fermentation *S. cerevisae* has two valuable uses in school genetics, namely for studying natural and induced mutation and for demonstrating complementation as evidence on how genes might work.

1 *Natural and induced mutations*

1.1 A haploid adenine requiring mutant (*ad* strain) that forms red colonies is particularly useful for this purpose since a back mutation to wild type, which forms white colonies, is readily recognisable.

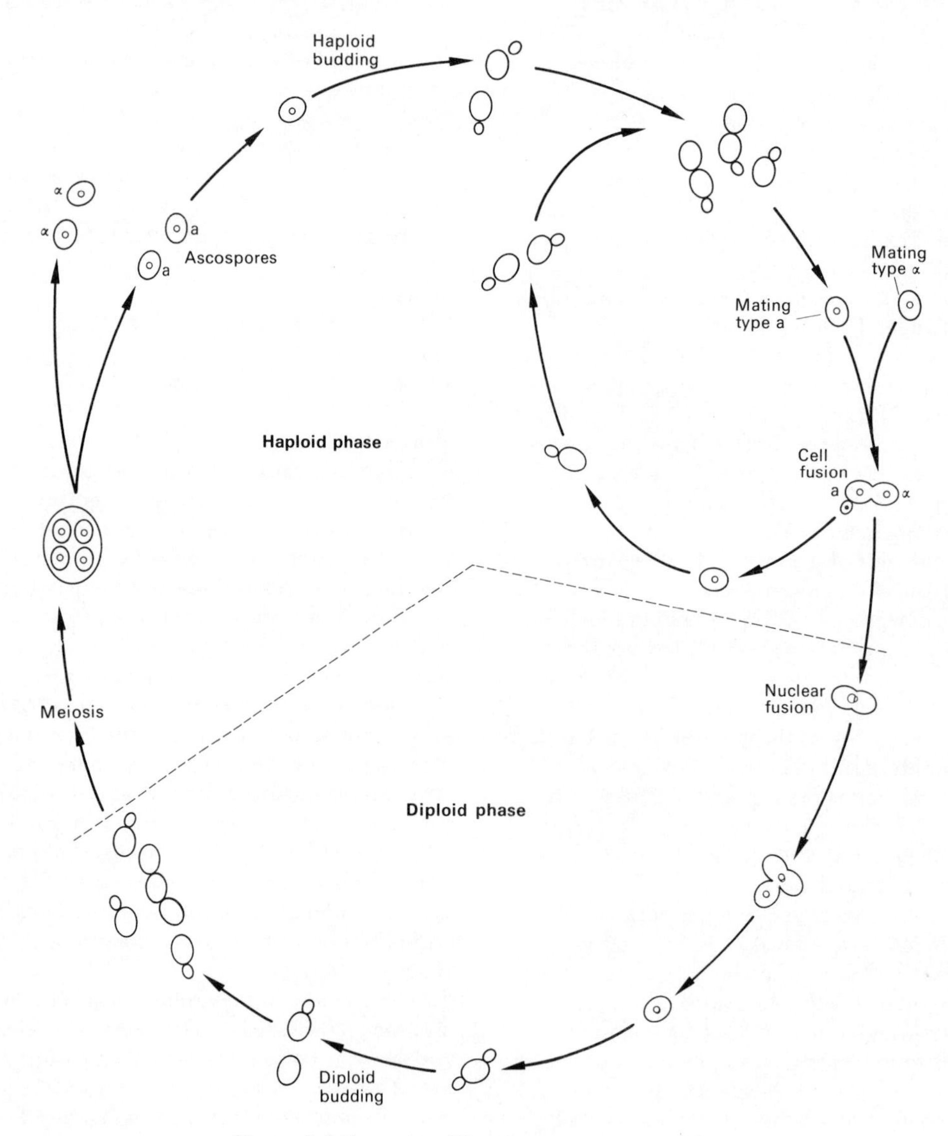

Figure 9 Life cycle of Saccharomyces cerevisiae

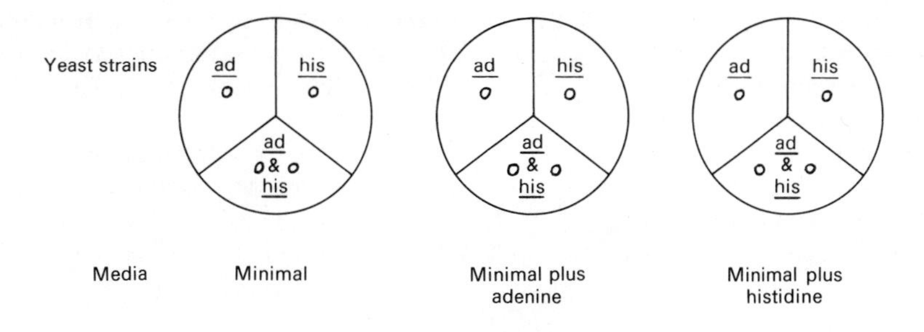

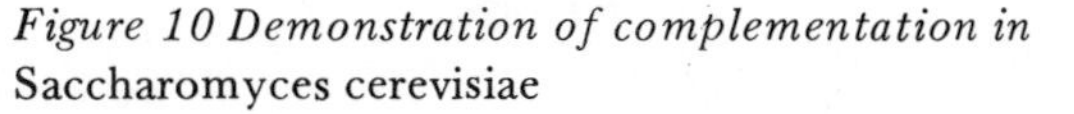

Figure 10 Demonstration of complementation in Saccharomyces cerevisiae

It should be shown first that the *ad* strain will only grow on minimal medium plus adenine.

Preliminary investigations using serial dilutions of cells shaken up in sterile distilled water should be carried out to determine suitable densities of cells for the purpose.

Plates are then streaked with measured volumes of the selected suspensions and together with a corresponding set of unstreaked controls are subjected to different growth conditions of temperature and light including ultra-violet radiation. The plates are then incubated for twenty-four hours at 30°C after which the ratios of white colonies to total colonies in each case are calculated.[13,14,16]

1.2 An alternative to 1.1 above makes use of the fact that many respiratory and nutritional mutants of yeast are stained differentially by the stain Magdala Red.[112] They stain bright red in contrast to wild type which stains only a pale pink so that a wide range of mutants can be identified as such even though their mutant characteristics are not known. In this case a prototrophic wild type strain is used. It is streaked as before but in complete medium with Magdala Red added. The investigation can be made quantitative by using ultra-violet radiation for measured periods of time and a curve of dosage against mutation rate drawn.

2 Complementation

Two haploid mutants which can be conveniently used for this demonstration are the *ad* strain used in 1 above and a white histidene requiring *his* strain.[105] White diploid colonies which can grow on minimal medium can be formed by placing a drop of a suspension of one mutant on a drop of suspension of the other so long as they are of opposite mating types. The demonstration can be set out as shown in fig. 10.

Sordaria fimicola *13*

Sordaria is similar in many ways to *Neurospora* but as it is less liable to give rise to contamination in the laboratory it is used here in preference to that organism. Both fungi belong to the group Pyrenomycetes of the Ascomycetes; these are characterised by long thin asci in which the ascospores are arranged linearly. Their great value is that the position of a pair of spores in the ascus indicates the pole of the cell to which its chromosomes moved at the first and second meiotic divisions. As mutants are available which affect spore colour it is possible visually to determine if and when a cross-over between the centromere and the locus for the mutation has occurred and thus to find the cross-over value and the chromosome map interval.

In *Sordaria,* and *Neurospora* also, the two meiotic divisions, which result in the formation of four cells in the ascus, are followed by a mitotic division so that eight ascospores are produced consisting of four pairs of spores, each pair being genetically similar. The life cycle of *Sordaria* is shown in Fig. 11.

To study the effects of crossing-over during meiosis two haploid strains of *Sordaria* are needed, one carrying the gene for the normal black spores, for example, Strain C7h, (AA) and the other the gene for white spores, for example, Strain C7h (1) (aa).

About four days before hybrid perithecia are required for classwork a petri dish of cornmeal agar is inoculated with the two strains of the fungus as shown in Fig. 12 and incubated at 28°C.

Sometimes, but not always, the strain with white spores seems to grow more slowly than the wild type and if this is found to be so from preliminary examination the problem can be met by inoculating the plate with the white-spored mutant one or two days before inoculating with the black-spored wild type.[14,16]

The perithecia will appear as small black spheres mainly on a line midway between the two points of inoculation. When hybrid perithecia are mounted in a

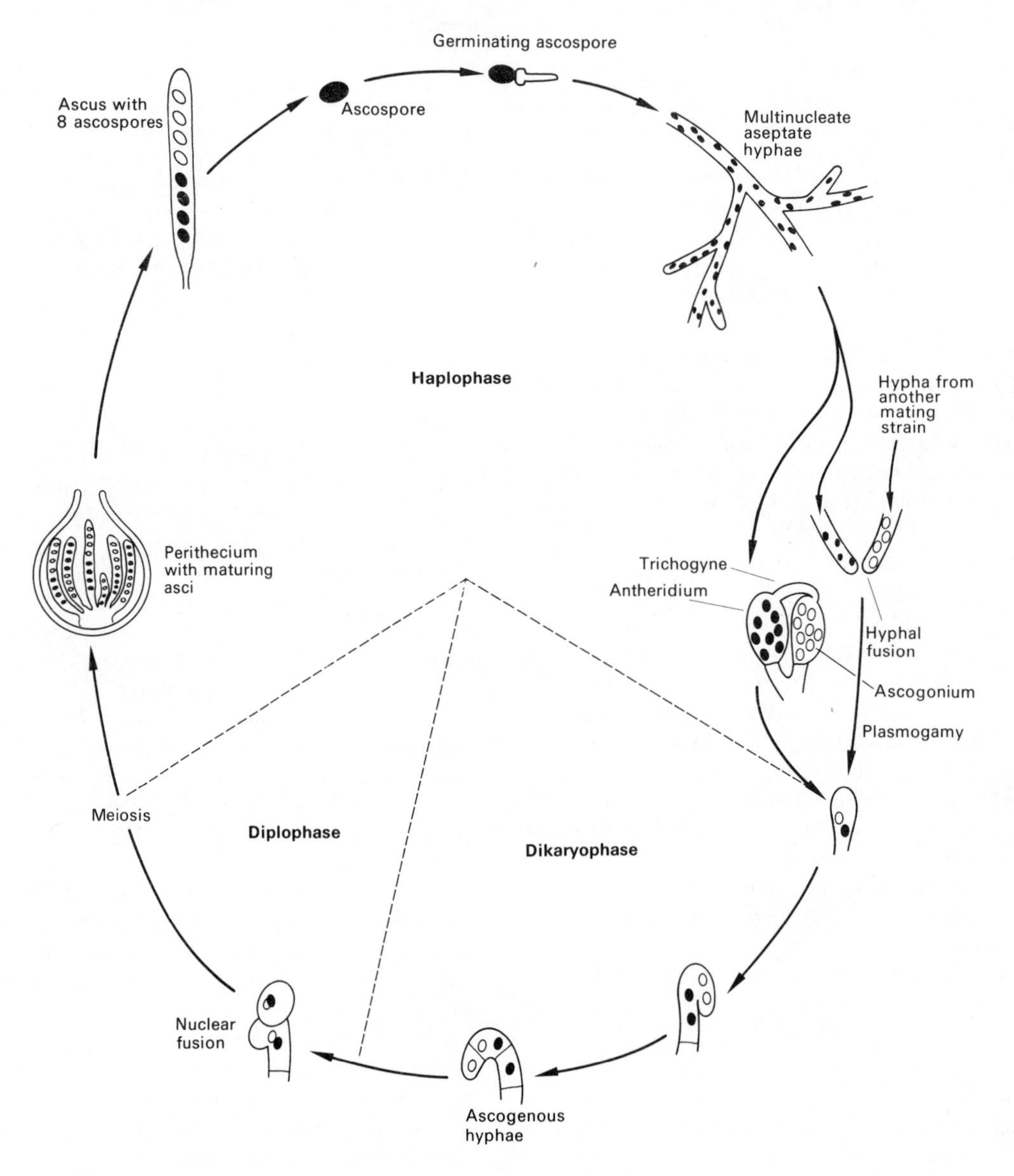

Figure 11 Life cycle of Sordaria fimicola

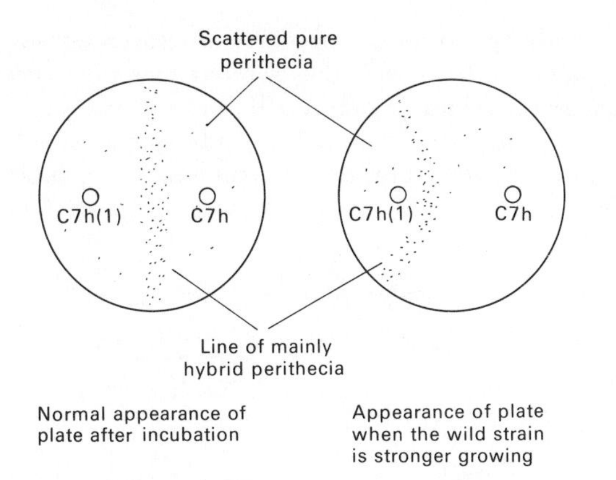

Figure 12 Sordaria-*appearance of plates*

drop of water and gently crushed by pressure applied to the coverslip the asci will be seen in microscopical examination to contain four black and four white spores each but in six different arrangements (see Fig. 13). Perithecia which are not hybrid have asci containing all black or all white spores. Sample counts of the six types of hybrid asci are made and the results from a class totalled. In one the frequencies obtained were:

A=164, B=168, C=77, D=68, E=83, F=59.

If A and B are statistically equivalent this indicates that the two centromeres orientate at random at the first division. If C, D, E and F are statistically equivalent this indicates that the two centromeres orientate at random at the second division.

For A and B: $\chi^2 = 0.05$ for 1 degree of freedom, $p = 0.87$.

For C, D, E, and F: $\chi^2 = 4.50$ for 3 degrees of freedom, $p = 0.28$.

From this analysis it follows that the centromeres do orientate at random in both cases.

The proportion of the total number of divisions in which a cross-over occurs between the centromere and the locus of the spore colour gene is a measure of the distance the gene locus lies from the centromere along the chromosome and thus a measure of map distance. In the case quoted above the value is given by $\frac{C+D+E+F}{A+B+C+D+E+F} = \frac{287}{619} = 46.4\%$.

But a difficulty arises from the fact that in *Sordaria* and other similar fungi we are concerned with the linear arrangement of the spores in an ascus and not, as in the case of test-crosses with organisms such as *Drosophila* and maize, with the actual genes carried by the four member chromatids of the tetrad, only two of which will have been involved in the

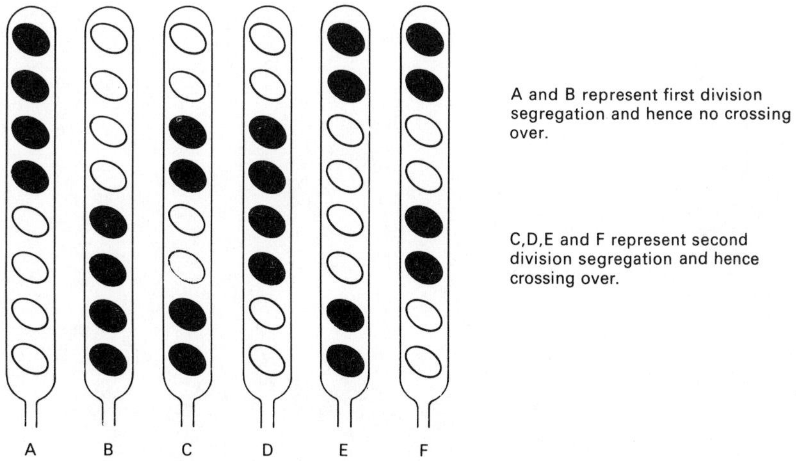

A B C D E F

A and B represent first division segregation and hence no crossing over.

C,D,E and F represent second division segregation and hence crossing over.

Figure 13 Sordaria-*hybrid perithecia*

crossing over. The result is that while all the asci resulting from a meiosis in which a cross-over occurred will show one of the arrangements C, D, E or F and are counted as recombinants only two out of every four chromatids that survive as gametes from meiotic divisions will carry a rearrangement of the genes concerned, the other two retaining the parental patterns. Therefore, in order to make cross-over values and map distances in these fungi comparable to those determined for higher organisms, the number of

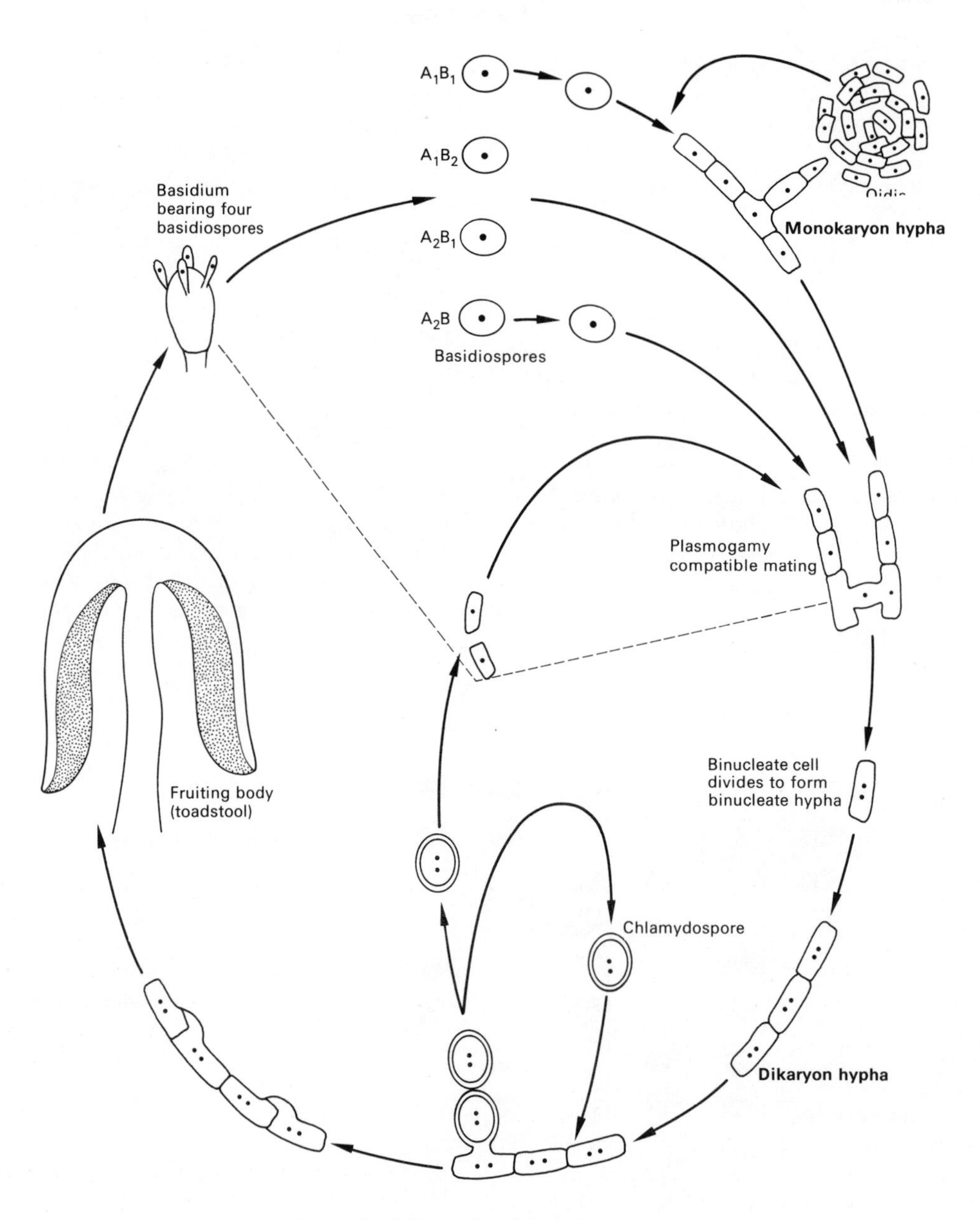

Figure 14 Life cycle of Coprinus lagopus

recombinants is halved. The map distance between the spore colour locus and the centromere from the data given above is therefore 23.2 units.

Coprinus lagopus *4, 10, 16, 17*

Coprinus lagopus is a Basidiomycete and one of the common Ink Cap Fungi. Part of its attraction as an organism for school work lies in the fact that besides its uses in genetics it illustrates the life history and mode of life of a common toadstool which is likely to be found on any dung heap or similar situation on a field expedition in autumn.[104]

Although *Coprinus* is especially useful for demonstrating complementation and biosynthetic pathways, which will be described here, it can also be used for fungal breeding experiments and for the induction and characterisation of mutants.

A knowledge of the life cycle is necessary before the procedures can be understood and this is shown in Fig. 14.

The basidiospores that are shed from the sterigmata of the gills give rise to monokaryon hyphae which have a number of different mating types based on two genetic loci, A and B. Dikaryons can only be formed by the fusion of hyphae of compatible mating types. Thus $A_5 B_5$ is compatible with $A_6 B_6$, but not $A_5 B_6$ or $A_6 B_5$. After the fusion of two compatible hyphae, each cell for a time contains two haploid nuclei and hence the term dikaryon. When two small pieces of agar containing two compatible monokaryon hyphae (each about the size of a match head) are placed side by side the successful formation of a dikaryon is indicated by the vigorous fluffy growth quite different from the less vigorous smooth appearance of a growing monokaryon.

1 *Complementation*

Mutant forms are known which cannot synthesise adenine (*ad*) or which cannot synthesise para-aminobenzoic acid (*paba*-1).

By the use of minimal media and media with additions of adenine or para-aminobenzoic acid these mutants can be identified. Furthermore complementation can be shown by growing the mutants together as shown in Fig. 15. Further information can be found in references 104 and 105.

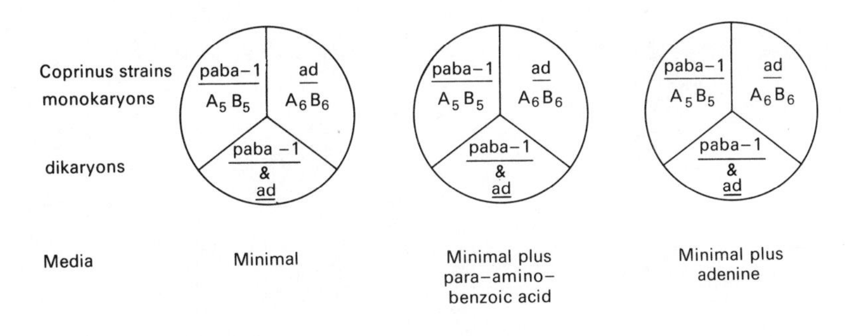

Figure 15 Demonstration of complementation in Coprinus lagopus

2 *Biosynthetic pathways*

Two choline requiring mutants of *Coprinus* are known, *chol-1* and *chol-2*, which differ in the point in the biosynthesis of choline at which the synthesis is blocked. This pathway consists of the successive replacements of hydrogen atoms of the amino group by methyl groups and the final addition of a further methyl group. The complete pathway is shown diagramatically in Fig. 16.

The *chol-1* mutant is blocked at Step 2 and the *chol-2* mutant at Step 3. It is possibly only a matter of time before a third mutant blocked at Step 1 is discovered. In the meantime the pathway can be demonstrated and room left for the third

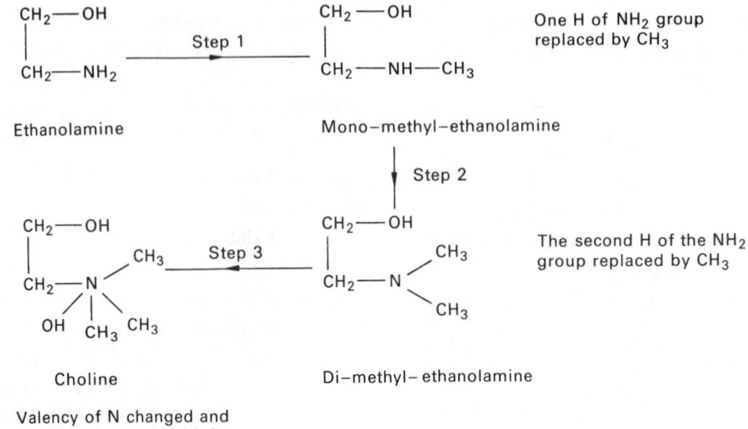

Figure 16 Synthesis of choline

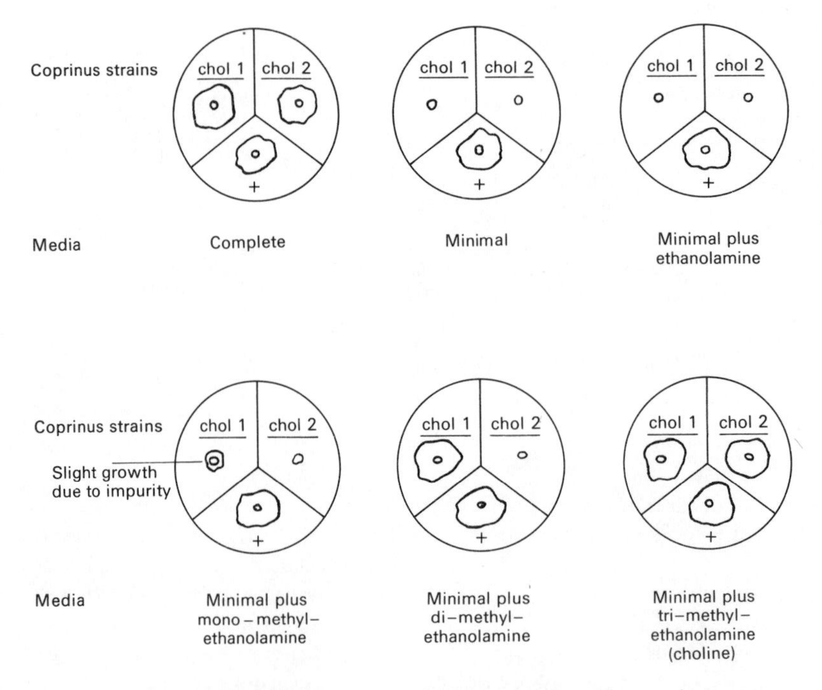

Figure 17 Demonstration of biosynthetic pathways in Coprinus lagopus

hypothetical mutant by setting out a demonstration as shown in Fig. 17 and incubating the plates at 30°C.

It may be found that *chol-1* will grow slightly on the minimal and monomethyl ethanolamine medium. This is probably due to small traces of dimethyl ethanolamine in the monomethyl ethanolamine, which illustrates the extreme sensitivity of these nutritional mutants. Both these choline mutants are vigorous growers, much more so than the wild type, once their nutritional deficiencies have been made good. The plates should therefore be examined every day and placed in a refrigerator until required once a good area of growth has been obtained.[14,16]

Aspergillus niger *2*

Aspergillus spp. have been used extensively in fungal genetics but only one species and one use are included here. These are *A. niger* and its reactions to the absence of traces of several metals, which are so sensitive that the organism can be used to detect such deficiencies in soils and plants growing on them. Such determinations are of economic importance as a number of diseases of crops and stock are due to trace metal deficiencies. For example, boron deficiency leads to 'heart rot' in mangolds and sugar beet, and cobalt deficiency to 'pining sickness' in sheep, while copper deficiency in the diet of ewes leads to the fatal 'swayback' in their lambs.

A problem associated with demonstrating trace element deficiencies is the need to use chemicals free from the element in question. For this reason magnesium deficiency is used for this demonstration as the result shows up even though small traces of magnesium may be present, that is, a threshold concentration of magnesium has not been reached.

Comparable samples of the fungus are inoculated on to two sets of plates and incubated at 37°C. One set of plates contains complete medium and the other complete medium minus magnesium. The fungus growing on the complete medium grows vigorously and produces black conidia and spores whereas on the magnesium deficient medium growth is retarded and the conidia and spores are pale brown.[15,17,113]

Fusarium culmorum *2*

Species of this important fungal disease organism of plants can be used to demonstrate environmental variation with different nutrients as the variable. Comparable pieces of *F. culmorum* hyphae are inoculated on to plates of Dox, malt and potato agar respectively and incubated at 36°C.

Bacteria

Bacillus subtilis and Sarcina lutea *2*

These two bacteria show the effect of the environment, in this case the amount of oxygen available, upon growth and thus demonstrate the interaction of inheritance and environment.

The procedure is similar to that described for *Aspergillus niger* but in this case suspensions of the bacteria are 'seeded' on to plates of nutrient and thioglycollate agar respectively and incubated at 35°C. The thioglycollate medium is anaerobic.

Appendix 1 Management

Mammals

Table 13 Management of small mammals

Management		Species	Mouse *Mus musculus*	Rabbit *Oryctolagus cuniculus*
Containers (Cages)	Type		Plastic 'shoebox' base and metal gridding top preferably stainless steel.	All metal preferably stainless steel or wooden hutch
	Suggested minimum dimensions of cage	Height cm	12.5	45.0
		Area cm^2	500.0	5500.0
	Number of adults per cage		Pair or trio	One medium size
	Litter		Granulated peat or clean white, softwood sawdust	As for mouse
	Nesting material		Good quality hay or clean wood wool	Good quality hay
	Cleaning		Twice a week. Cages to be disinfected.	At least once a week. Cages to be disinfected.
Environment (External to cages).	Temperature °C		20.0–22.0	16.0–20.0
	Humidity % relative		50.0–55.0	40.0–45.0
Nutrition (it is assumed that water is freely available from suitable bottles and pellcted food from suitable hoppers)			Pellet diet FFG(M) May be supplemented by small quantities of fresh food	Pellet diet SGI or 18 Supplemented by fresh food e.g. carrot
Handling			Lift by grasping the base of the tail	Lift by grasping the neck and shoulders whilst supporting the rump
Sexing			Distance between the anus and the urino genital opening in the adult male is about one and a half to twice that in the female. Nipples of lactating female visible.	Male genital opening circular in appearance and penis can be extruded. Female opening I shaped.

Guinea-pig or cavy *Cavia porcellus*	Syrian or golden hamster *Mesocricetus auratus*	Rat *Rattus norvegicus*	Mongolian gerbil *Meriones unguiculatus*
As mouse. Alternatively keep in a large pen with sides 0.5m high, or in hutch.	As mouse.	As mouse.	As mouse.
30.0	20.0	25.0	20.0
2500.0	1000.0	1000.0	1000.0
Pair	Pair or single	Pair	Pair
As for mouse	As for mouse	As for mouse	As for mouse
Good quality hay	As for mouse	As for mouse	As for mouse
As for rabbit	As for mouse	As for mouse	As for rabbit
16.0–20.0 50.0–60.0	20.0–22.0 45.0–50.0	18.0–22.0 45.0–50.0	18.0–22.0 45.0–55.0
Pellet diet RGP or SGIV supplemented by fresh food e.g. carrot	As for mouse	As for mouse	As for mouse
As for rabbit	Lifting by cupping in both hands	As for rabbit	As for Syrian hamster
Male as for rabbit. Female opening Y shaped. (Note that both female and male have prominent nipples.)	As for mouse	As for mouse	As for mouse

Table 13 (contd.)

Management		Species	Mouse *Mus musculus*	Rabbit *Oryctolagus cuniculus*
Breeding	Recommended minimum breeding age	Weeks	10	24
	System		Trios (2♀:1♂) or pairs established shortly after weaning	Take female to male for mating
	Handling young		May be handled shortly after birth	Best not to handle till at least one month old
	Recommended maximum breeding age.	Months	12	24
	Gestation period.	Days	19.0–21.0	30.0–32.0
	Usual number in litter		8–11	4–6
	First oestrus cycle after birth. Days. (This is an indication of when the next mating can or should take place.)		Shortly after	35
References	General 13, 34, 36, 62, 68		19, 20, 26, 38, 41, 49	35, 59, 67

Fish

Guppy Poecilia reticulata (= Lebistes reticulatus)

Container Aquarium tank allowing 65 cm^2 surface area per 2.5 cm fish (less tail fin). Plastic aquaria 30 x 20 x 20 cm (12 x 8 x 8 inch) are very suitable and can house six to eight fish. Females can be isolated in 500 cm^3 beakers with an internal diameter of 8–10 cm. Agitation of the water using an aeration block increases the holding capacity.

Environment Tap water is suitable if allowed to stand before use. Temperature of 20–25°C. (Range 15–30°C; breeding usually will not take place when below 20°C). Maintain temperature using electric heater and thermostat suitably earthed and protected by a fused circuit.

Nutrition Omnivorous Feed with a variety of foods including dried and live for example *Daphnia* the water flea, *Artemia salina* the brine shrimp, *Tubifex spp.* and white worms. Avoid overfeeding. The fry need to be fed on infusoria or other small live food several times a day.

Sexing Male is smaller than the female, brightly coloured and has a sharply pointed modified fin or gonopodium. Adult male about 2.5 cm, female 5–6 cm in length.

Guinea-pig or cavy *Cavia porcellus*	Syrian or golden hamster *Mesocricetus auratus*	Rat *Rattus norvegicus*	Mongolian gerbil *Meriones unguiculatus*
16	10	12	12
Pairs if in cage or hutch. Harem (one male to several females) if in pens	Pairs can be established after weaning. Otherwise bring female to male for mating.	Pairs	Pairs
As for mouse	Best not to handle till 16–17 days old	Best not to handle till at least one week old	As for mouse
18–24	12	12–15	18
59.0–73.0	15.0–17.0	20.0–23.0	25.0–28.0
3–5	5–7	9–11	4–6
Immediately after	1–8	As for guinea-pig	As for guinea-pig
64			

Life history Live bearers, sexually mature at two to three months, fully grown at six months. Gestation period four to six weeks depending upon the food, temperature, age of the female, and other environmental conditions (about three weeks at 25°C). Between twenty to one hundred young born at a time.

Breeding Since the sperm is retained within the female for some time it is essential to separate the sexes as early as possible. Place the female in a 'separator' before the young are born to avoid cannibalism. Breeding will take place all the year round if the temperature and illumination are adequate.

References 40, 50, 100.

Birds

Table 14 Management of birds

Management	Species	Budgerigar *Melopsittacus undulatus*	Canary *Serinus canarius*	Domestic Fowl *Gallus domesticus*
Containers	Type	Breeding cage or aviary	Breeding cage (double) or aviary	Deep litter about 2700 to 3600cm^2 per bird on wire floors about 2250cm^2 Movable runs, if space available, allowing at least 5000cm^2 per bird.
Suggested dimensions for breeding	Length Height Depth	As canary with partition removed.	65–100cm (25–40 in) 45cm (18 in) 30cm (12 in)	
	Perches	as Canary	15mm diam (Preferably oval about 15 x 12mm)	
	Litter	Sand tray	Sand tray	Hay/Woodwool
	Nesting	Nest box. 15 x 15cm (6 x 6in) and 23cm high (9 in) Concave interior; base about 14cm (5.5 in) diameter. Entrance hole 4cm diam (1.5 in) Inspection lid. Fit nest boxes as high as possible.	Wooden pans about 12.5 x 12.5cm (5 x 5in) and 5cm (2 in) deep	
Environment	Temperature	Protect from cold. Minimum indoor of 10°C (15°C when young)	As Budgerigar	As Budgerigar
Nutrition (Food dispensed in suitable hoppers. Water freely available).		Mixed seed such as millet, canary grass. Supplemented by fresh food e.g. plantain, salad vegetables and as for canary.	Two to three parts by weight of canary seed and one part of red rape. Supplement with fresh food and high protein food e.g. egg for hen. Proprietary brands for enhancing the red colour are available.	Proprietary mashes, crumbs or pellets.
Handling		Neck between first and second fingers, back along palm of hand, thumb and first finger encasing the body. Wings folded against body.	As for Budgerigar	Wings folded against body. Lift gently.
Sexing		From about 12 weeks the upper part of the beak (the ceres) is blue in the male and brown in the female.	Difficult. Adult cock more thickset, active and bolder with strong song. (See 51 and 68)	Combs of cockerels red, tails stumpy and rounded.
Identification		Ringing or banding the legs	As for Budgerigar	As for Budgerigar

Management / Species		Budgerigar *Melopsittacus undulatus*	Canary *Serinus canarius*	Domestic Fowl *Gallus domesticus*
Life cycle	Sexual maturity	10–12 months	12 months	6–8 months
	Eggs laid	5–6 per clutch	1–6 per clutch laid at 24 hour intervals	280–300 per year
	Incubation period	18 days	16 days	21 days. for artificial incubation practice see references
Breeding		Separate sexes in winter. Pair in February or March	Hens housed together, cocks separately before breeding season. Pair in February or March	Allow one cock to 10–12 hens.
References	General 62, 67, 68	72, 73	51	67

Invertebrate animals

Drosophila

Obtaining virgin females

All experimental crosses must start with virgin females. To achieve this, advantage is taken of the fact that females do not mate until at least ten hours after emergence from the pupa. Under natural conditions emergence occurs most frequently in the early morning, that is, when it becomes light. This condition can be simulated by switching on a light in the incubator or simply leaving the incubator door open. Then, if the stock cultures are cleared as early as possible in the morning and all the newly emerged flies collected as late as possible in the afternoon all the females will be virgin. The sexes are then kept separate until they are needed (Fig. 18). Two checks on the virginity of females are possible if any doubt exists. Virgin flies are paler in colour than other females and a dark faecal pellet can be seen through the translucent wall of the abdomen. If any of the female flies transferred from the stock bottles to await crossing are not virgin, larvae will appear in the tube to which they have been put one or two days later.

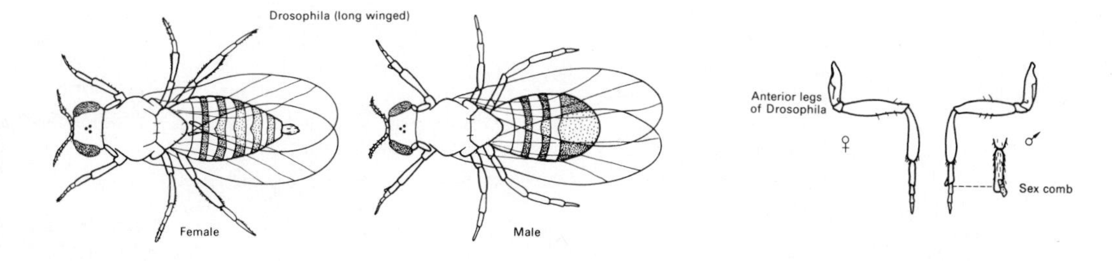

Figure 18 Male and female Drosophila *with anterior legs*

Setting up crosses
Have tubes ready with labels giving details of the cross to be made. Anaesthetise the flies (see Fig. 19) and select about five males and five virgin females of the required strains. Put selected flies into the tubes and enter details into the record book. Place tubes in incubator set at 25°C. After three days the adults may be transferred to a second tube to avoid overcrowding. On the eighth day, remove the adults to avoid any overlapping with the next generation. Counting will begin on the 14th day and should continue for a further two weeks, all counted flies being either destroyed in 70% alcohol or stored in a fresh container. There are differences in the average development rates of males and females and of different strains so that counting should continue until no more flies emerge.

A programme for say the first year in the sixth form would be to carry out monohybrid and dihybrid crosses simultaneously and to determine the ratios in the F_2 and backcross progenies and that each group of four students would be responsible for a complete set of experiments and results. There remains the decision as to whether reciprocal crosses are made in each case and whether both ways of combining the characters in the dihybrid cross are to be used. To simplify matters and to keep costs low let us decide to make reciprocal crosses for the monohybrid experiment but not for the dihybrid one and to use the combination of wild type with each character separately for the dihybrid. It will still be necessary to produce the double recessive homozygote for the backcross if this stock is not kept. This can be done during the experiments. There will be three stages each of about two weeks.

Stage 1. Building up the stocks of flies (virgin females and males) required for the initial crosses.
Stage 2. Making up the initial crosses and producing the F_1 progenies and the virgin flies required for the backcrosses. The F_1 x F_1 crosses will not, of course, require virgin females.
Stage 3. Making up the F_1 crosses and backcrosses, producing the progenies and scoring results.

The collection of virgin flies is perhaps the most troublesome part of the proceedings and that is why the backcrosses are often omitted, but something is lost in doing this especially as an understanding of the backcross ratios is essential for the later studies of linkage and three point crosses. So as to facilitate the collection of virgin females in the three days between the 12th when the first flies begin to emerge and the 15th when the first crosses are made, it is convenient to make the crosses on a Friday so that three whole school days are available for this purpose.

If each group of four pupils work as two pairs A and B, the schedule would be something like the one given in Table 15. Tubes of medium can be made up in large batches and kept in a refrigerator.

Table 15 Schedule for Drosophila *crosses*

Stage	Day	Teacher or Technician to prepare for each group of 4	Students	Notes
Preparation	0 Thurs	4 sterile tubes with medium and ready for use		
	1 Fri	Confirm that vge/vge stocks for double recessive are available or set up the necessary cross	A one sub-culture each of +/+ and e/e B. one sub-culture each of e/e and vg/vg	Use flies from stock bottles Care taken that no flies escape into the room
	5 Tues		A and B. Remove parents from all sub-cultures	Parents may be transferred to new cultures; otherwise destroy
	11 Mon	09.00 Empty all culture tubes 8 tubes ready for virgin females and males	16.00 A and B. Label tubes and begin collection of virgin females and males	Emergence and collection of parents begins

Table 15 (contd.)

Stage	Day	Teacher or Technician to prepare for each group of 4	Students	Notes
	12 Tues	09.00 Empty all culture tubes	16.00 A and B. Collect virgin females and males	
	13 Wed	09.00 Empty all culture tubes	16.00 as day 12	
	14 Thurs	4 tubes ready for crosses	16.00 as day 12	
Crosses for F_1	15 Fri		Label tubes and make up crosses A. +/+ x e/e e/e x +/+ B. vg, +/vg, + x +, e/+, e	
	19 Tues		A and B. Remove parents from all experimental crosses	Parents transferred to fresh tubes or destroyed
	25 Mon	09.00 Empty all experimental crosses and a stock culture of e/e, vg/vg and vg, e/vg, e. 12 tubes ready for collection of virgin females and males	16.00 A and B. Collect virgin females and males from experimental crosses and cultures	
	26 Tues	09.00 Empty tubes and cultures	16.00 Collect virgin females and males from tubes and cultures	
	27 Weds	09.00 Empty tubes and cultures	16.00 As day 26	
	28 Thurs	09.00 Empty tubes and cultures 6 Tubes ready for crosses	16.00 As day 26	
Crosses for F_2 and Backcross	29 Fri		Label tubes and make up crosses A/F_2 +/e x +/e, e/+ x e/+ Back +/e x e/e, e/+ x e/e B/F_2 vg, e/++ x vg, e/++ Back vg, e/++ x vg, e/vg, e	
	33 Tues		A and B. Remove parents from all crosses	Parents may be transferred to new tubes to increase size of progenies
	39 Mon			Emergence of F_2 and backcross progenies begins
	43 Fri		A and B. Count all progenies to date and destroy	
	46 Mon		A and B. As day 43	
	50 Fri		A and B. As day 43	
			Continue until no more emerge	

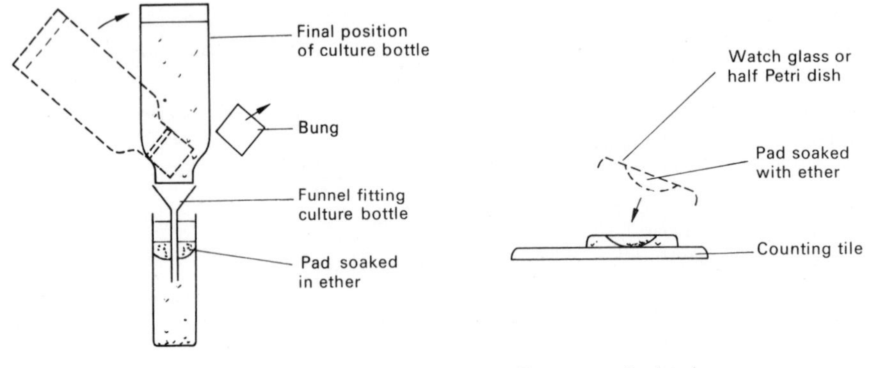

Figure 19 Drosophila *etherisers*

Setting up population studies
Set up three or more population cages (Fig. 20) with equal numbers of males and females e.g.

1 40 ebony body (or vestigial wing)
2 30 ebony body (or vestigial wing): 10 wild type
3 20 ebony body (or vestigial wing): 20 wild type
4 10 ebony body (or vestigial wing): 30 wild type
5 40 wild type

Each week replace one tube by a fresh one in rotation. The flies that hatch from the tubes that have been removed can be returned to the main population after a further week and then the tubes discarded. For counting, the populations are anaesthetised lightly by replacing the cotton wool plug in the centre hole with one soaked in ether. They are then counted and returned either to the same or a fresh population chamber. There is usually no need to count males and females separately, only the total number of each phenotype.

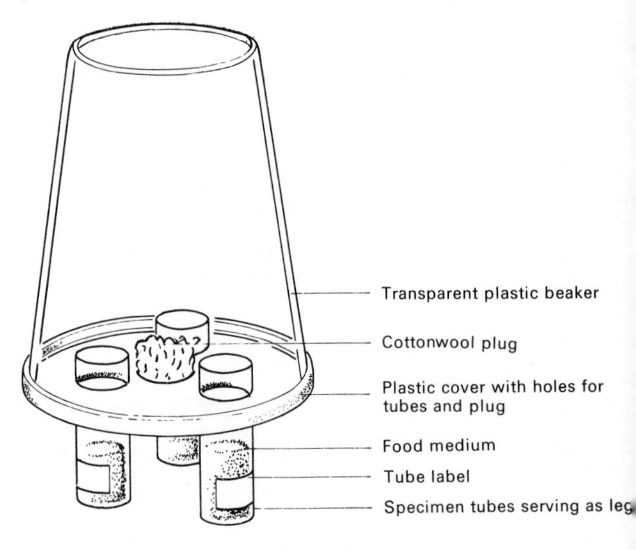

Figure 20 Drosophila *population cage*

Tribolium

Setting up crosses. Sieve out and sex sufficient pupae from the appropriate stocks. For each cross transfer two males and two females of the parental types into a tube half full of the food medium and incubate at 30°C and 70% relative humidity.

Examine after a week to ten days to confirm that the adults have emerged. After two more weeks remove all four parents.

From then on examine once a week and collect sufficient pupae for the F_2 and if these are to be made, the backcross matings. For the backcrosses the pupae must of course be sexed and the appropriate

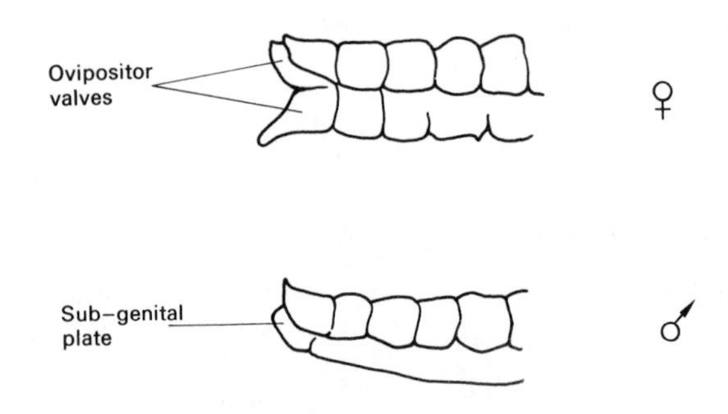

Figure 21 Terminal region of the abdomen of Chorthippus *spp.*

matings made. For the F_2 matings sexing is not necessary so long as a sufficient number of pupae are used to ensure at least some individuals of each sex. The remainder of the F_1 progeny are best destroyed.

The same general procedure is used for the F_2 and backcross generations as for the F_1 but this time the counts are made at weekly intervals as soon as the adults begin to emerge. Once counted an adult must be removed. Counting should go on for as long as living larvae and pupae remain, but not longer than four weeks after the first adults emerged, in case some mating and egg laying occurred before the earliest adults to emerge have been counted and removed.

Table 16 Management of invertebrate animals

Management	**Species**	Vinegar Fly *Drosophila melanogaster*	Flour-beetle *Tribolium confusum*	Locusts *Locusta migratoria (Schistocerca gregaria)*	Field grasshopper *Chorthippus spp. (C. bruneus)*
Containers	Type	Any suitable e.g. small milk bottles or glass tubes. For genetic crosses it is convenient to use 7.5 x 2.5cm glass specimen tube closed with a foam plastic bung.	Any suitable e.g. jam-jars closed with fine muslin or nylon net or as for *Drosophila*.	Cage approximately 50 x 40 x 40cm with false floor under which aluminium cylinders or glass specimen tubes may be placed for egg laying.	As for locusts or large glass containers or aquaria
	Litter	–	–	–	Sand on floor of aquaria
Environment	Temperature	Basic stocks at 19–21°C Experimental stocks and crosses at 25°C	25–30°C	Daytime 32°C Night-time 28°C	25°C
	Ventilation	–	–	Air movement through cages essential	Some circulation desirable to maintain humidity
Humidity % relative		High within tubes but condensation must be avoided	60–70	Low	Low
Nutrition		Food medium and yeast (see page 111)	Flour and yeast medium (see page 112)	Fresh uncontaminated grass and protein supplement (see page 111). Some water may be necessary.	Fresh uncontaminated grass
Handling		Flies have to be anaesthetised to handle. It is advisable to sort them on a light background e.g. a white tile.	Remove from medium by sieving. Sort on light background. For handling large numbers a pooter is useful.	Grasp thorax gently with wings folded along body.	If necessary as for locusts.

Table 16 (contd.)

Management / Species		Vinegar Fly *Drosophila melanogaster*	Flour-beetle *Tribolium confusum*	Locusts *Locusta migratoria* *(Schistocerca gregaria)*	Field grasshopper *Chorthippus spp.* *(C. bruneus)*
Sexing		Male fly has shortened rounded abdomen appearing black at the rear and sex comb on fore-legs (see fig. 18)	Pupae of female have strongly protuberant and diverging genital appendages.	Ovipositors of adult female clearly visible, the tip of the abdomen appearing divided horizontally.	As for locust. Adult males have smaller bodies and longer wings which protrude beyond the end of the abdomen (see fig. 21)
Life cycle	Egg	Laid on surface of food medium, hatches in about 20 hours if kept at 26°C	Laid in food medium. Give large volume of this per adult to avoid cannibalisation of eggs. Eggs hatch in about 7 days if kept at 25°C	Laid in moist sand. Eggs hatch in about two weeks if kept at 30°C	Laid in moist sand. To break diapause first keep egg pods in contact with water at 25°C for about two weeks, then chill to 5 ± 2°C for nine weeks. Return to 25°C and nymphs should emerge in about fourteen days
	Larva	Burrows into food. Moults twice, pupates in about three days if kept at 25°C.	Pupates in twenty-five to forty days if kept at 25°C	— Nymphs with five instars; emergence to final instar about twenty days if kept at 30°C	— Nymphs with five instars.
	Pupa	Larva pupates on side of container or paper insert. Imago emerges in about five days if kept at 25°C.	Larva pupates in food medium. Imago emerges in about five or seven days if kept at 25–30°C.		
	Imago	Virgin females essential for genetic crosses			
References		32, 33, 61, 74	46, 47	28, 43, 48	53

Table 17 Management of plants

Organisms / **Management**	Greenhouse Plants	Half-hardy annuals	Hardy annuals	Biennials	Hardy perennials	Hardy perennials (bulbous)
	Tomato *Pelargonium*	Snapdragon Stocks Maize	Cereals Pea Sweet pea	Beets Brassicas Wallflower	Roses *Tradescantia* *Trifolium*	*Allium* *Crocus*
Container or planting position	Pots. May be placed in the open in the summer or planted out	Seed trays then planted out Snapdragon and stocks can be grown in pots	Outdoors Can be grown in large pots	Outdoors	Outdoors Can be grown in large tubs, *Trifolium* in seed trays	Outdoors Can be grown in pots
Environment						
Temperature	Frost free minimum 5 °C	Frost free minimum 5 °C	–	–	May need protection from frost	–
Compost or soil	Sow in John Innes seed compost then pot on into 1, 2 and 3	Sow in John Innes Seed compost	Good loam	Good loam	Good loam	Good loam rich in humus
Aspect	Greenhouse may need shading to control temperature in summer	Full sun	Full sun	Full sun	Full sun Primulas need shade	Some shade
Propagation (Notes on cross-pollination are on pps 102-104	Seeds: (Tomato) Sown Jan/Feb 20–25 °C lower temperature after germination Remove lateral shoots from tomato. Pinch out tip of plant after two or three trusses of fruit have formed.	Seeds: Sow thinly in seed tray in February at 15 °C Prick out into boxes or peat pots soon as convenient. Keep at 10 °C. Harden off in cold frame and plant out in mid-May.	Seeds: sow about 5cm below surface in shallow drill, Sept/Nov for winter cereals; Feb/April for spring varieties; Feb/March for peas. Sweet pea seeds may need the seed coat cracking before sowing. Cereals may need protective cages to prevent access by birds.	Seeds: sow March/April and over winter as small plant	Budding: Roses Division of plants for herbaceous forms	Separation of new grown bulbs or corms
Cuttings: (*Pelargonium*) shoot 8 to 10 cm severed below leaf joint. Insert in well drained compost at 10–15 °C						
References General 90, 97	86				95, 102	

Crossing plants

Techniques for crossing any plants are essentially similar, the detailed notes given for a few species describe the basic techniques.

1 *Cereals*

Emasculation

Emasculation has to be carried out before the anthers dehisce and emerge from between the lemma and the palea on flowering. It becomes necessary therefore to time the operation fairly precisely. The first indication is given when the lower spikelets of some of the ears flower, that is the stamens and stigmas are exserted, usually about the middle of June. Suitable ears are then selected and the upper and lower spikelets removed with fine forceps leaving four or five spikelets in the middle of the inflorescence. The two glumes of a selected spikelet are then gently separated by means of fine forceps to expose the contained florets. All but one fertile or hermaphrodite floret are removed and it is this remaining floret that is emasculated. In barley (*Hordeum*) it is only the middle spikelet in each group of three that is hermaphrodite and it only contains one floret so that barley is probably the easiest of the cereals to emasculate and cross. Wheat has one or two hermaphrodite flowers and a number of male or sterile flowers in each spikelet, oats two or three, and rye two. After the unwanted spikelets and florets have been removed, the remaining hermaphrodite floret is opened by separating the lemma from the palea by means of very fine forceps and the three stamens pinched off with the forceps before closing the lemma and palea together again. It is good practice to place the stamens on the thumb nail of the hand steadying the ear as they are removed to ensure that three are taken from each floret and that none of them is ruptured.

Pollination

Whole anthers from the male parent can be placed inside the floret either when emasculation has been completed or a day or two later. The ear, now with only a few emasculated and pollinated florets, should be enclosed in a cellophane bag and a label attached giving details of the cross that has been made. It is conventional to give the female parent first. A note should also be made in the record book.

Maintaining stocks of Xantha barley

The seeds provided as Xantha strain are produced by the self-fertilisation of heterozygous plants and give on average three phenotypically green plants to one yellow plant. The yellow plants die when the reserve food in the seed is used up, generally after three or four weeks. Of the green plants remaining, two out of three are heterozygous, and as barley is normally self-fertilised, the seed from these will, in its turn, segregate to give three green plants to one yellow mutant. The problem thus resolves itself into how to identify the heterozygous plants. In the case of barley, and a high proportion of the other plants with lethal chlorophyll mutants, the heterozygotes cannot be distinguished from the green homozygotes so that the required heterozygotes have to be identified by test sowings of the seeds they produce.

The surviving green plants are grown to maturity and the seed from each harvested into packets separately. Test sowings of small samples from each plant are made; those yielding green plants only are classified as homozygotes, which incidentally provide the most suitable seed for the normal variety required in the barley demonstration (p. 62.2). Those yielding some yellow plants are heterozygotes and the remaining seed from these can be bulked for future use.

2 *Edible pea*

As peas are normally self-pollinated the F_1 generation will self naturally.

Emasculation and pollination

The female or seed parent has to be emasculated before the anthers have released any pollen so the time has to be chosen when the buds are as large and as easy to handle as possible, but before the anthers dehisce too easily. Fortunately buds at different stages of development can usually be found together and a convenient choice made. Varieties differ in the point when they are ready but as a rough guide a bud 5–8mm long will often be found to be suitable. (When a bud or buds have been selected the remaining flowers or buds on the stalk should be removed). The emasculation is best carried out with a fine pair of forceps. The first step is to open the bud and expose the ovary with its surrounding cylinder of ten stamens. This is done by gently lifting the standard and moving aside the two wings by means of the forceps and then holding them apart by the index finger of the hand holding the bud. The bottom of the keel can then be slit by using one of the sharp points of the forceps. This allows one side of the keel to be turned back to expose the ovary. The anthers are free at the anterior end of the flower and should be picked off as far back along the filaments as possible to prevent rupturing the anthers and inadvertently allowing pollen to get on to the stigma. It is important to check that all the anthers have been removed by examining the bud finally by means of a hand lens. Pollination can be carried out at once. This may be done by collecting pollen from the pollen parent on the roughly torn edge of a piece of filter paper and dusting it on to the stigma, or by using a dehisced stamen in the same way. Pollination by camel hair brush is often advocated and this is suitable so long as the same pollen parent is being used but the brush must be sterilised, or changed, when a new pollen parent is to be used.

After the pollination is completed the bud(s) should be enclosed in a cellophane bag and labelled.

3 *Maize*

Maize is wind pollinated, the male inflorescences or tassels occurring at the top of the stem and the female lower down. Emasculation is by the total removal of the male inflorescences.

If cobs from the experiments described above are required as museum specimens, care must be taken when drying the ripening cobs in the autumn as they seldom dry off naturally in Britain. A good plan is to cut off the cobs as soon as the grains are fully formed and to peel back the bracts that enclose the cobs. The cobs should then be allowed to dry fairly slowly in a dry, well-ventilated place. This can be done very effectively by impaling the cobs on long thin nails driven through a piece of thin wooden board.

4 *Snapdragon*

Most varieties are self-fertile and the anthers dehisce and the stigmas become receptive at about the same time so that the flowers have to be emasculated before any pollen is shed. Varieties vary to some extent in the stage at which the anthers dehisce but the appropriate time can soon be determined by inspection; it is often just before the lips of the flower finally unfold but may be two or three days earlier. The flowers are in spikes of about a dozen or more flowers so that a spike with about four buds at about the right stage is chosen and all the other flowers are removed and the axis carrying the smaller buds pinched off.

Emasculation

By holding the bud between thumb and forefinger of one hand and pressing laterally gently, the lips can be put under tension and then folded back by means of the points of a fine pair of forceps held firmly together. The upper lip is then held back by the forefinger and the four anthers pinched off by means of the forceps. It is wise to count the stamens that have been picked off as a double check and to make sure that no pollen has been shed from them. The lips of the flower are then returned to their normal position and when all the buds have been emasculated the spike is enclosed in a cellophane bag and labelled.

Pollination

Two or three days after emasculation the flowers will be ready for pollination. Probably the best way to do this is to take a ripe stamen from a newly opened flower of the selected pollen parent and holding it in forceps, touch the stigma of the emasculated flower of the seed parent with it in such a way that it is well dusted with pollen. The torn edge of filter paper dusted with pollen can also be used. When all the flowers remaining on the spike have been pollinated, the spike is once again enclosed in the cellophane bag and the label completed.

Subsequent treatment

Even if no leaves have been enclosed in the bag, water from transpiration tends to collect in the bag. It is a good idea to remove the bag occasionally and shake out the drops. When the capsule is well formed and the petals about to fall off, the bag can either be removed altogether or the corners cut off in order to allow the fruit to ripen naturally.

When the capsules are nearly ripe they should be removed from the plant and put in envelopes together with their label. Each successfully pollinated flower produces more than a hundred seeds so that good F_1 and F_2 progenies can be grown in the following years.

Although not absolutely necessary it is usually best to carry out the F_1 pollinations in the same way but in this case of course prior emasculation is not needed. It is essential however to bag the flowers to be pollinated before they open in order to avoid cross pollination by visiting insects.

5 *Tomato*

The tomato is normally self-pollinated and in-breeding.

Emasculation and pollination

The five anthers which open inwards form a cone completely enclosing the style and stigma. When a cross is to be made emasculation has therefore to be carried out before the anthers dehisce and to be on the safe side this should be done a day or two before the flower opens. If all the buds on the truss are either removed or emasculated the whole truss can be enclosed in the same cellophane bag and appropriately labelled. Pollination can be carried out two or three days after emasculation. When F_1 plants are being selfed to produce an F_2 progeny, the self-pollination can be encouraged by gently tapping the plants or their supports.

Subsequent treatment

There are two methods for perpetuating the thiamineless mutant (see page 52).

a Homozygotes can be brought to maturity when treated with thiamine sprays and then allowed to self.

b Heterozygotes in an F_2 progeny can be identified by their slight chlorosis and then selfed to give a new F_2 type progeny.

When the fruits are nearly ripe they can be picked and laid out into labelled boxes until they are over-ripe when the skins can be removed and the pulp roughly separated from the seeds by rubbing through a sieve of suitable mesh size. A final separation can be made by washing off the rest of the pulp with water, after which the seeds are dried and stored.

Fungi and Bacteria

(see also references 106, 107, 108, and 109).

Studying genetics with micro-organisms requires a certain amount of basic equipment, an autoclave or pressure cooker, an incubator in the range 15°–35°C, a refrigerator and an ultra violet lamp besides ordinary laboratory glassware, petri dishes, microscopes, etc. The chief skill to be acquired is that of sterile technique in making crosses or for plating out. A transfer chamber is useful for this purpose especially if the work has to be done in a general purpose laboratory where contamination is always a problem.

Media

Details of the production of suitable media are given in Appendix 2 pages 108 to 112. Media and cultures may best be contained in suitable bottles eg McCartney, Universal or 'medical flats', or in petri dishes. Pre-sterilised plastic petri dishes are most convenient to use. Other containers, and all media, must be sterilised by autoclaving at 121°C ($103.5 kN/m^2 \triangleq 15 lbf/in^2$) for fifteen to twenty minutes.

Pouring media

When bottles or dishes are ready the medium prepared in large bottles or flasks is allowed to cool to about 50°C or heated to this temperature. The caps and mouths of the bottles and other containers are passed rapidly through a flame and the medium poured out and allowed to set. The lids of dishes should be lifted only sufficiently far to permit the medium to be poured in, thereby reducing the chance of airborne contamination. It is best if this operation and those of inoculating the cultures are carried out in a transfer cabinet which can be sterilised by ultra-violet radiation before use. A cabinet suitable for schools is now available commercially. If the bottle is rested at an angle whilst the medium cools a 'slope' is formed on which the culture may be grown.

Inoculating
Nichrome wire (20 SWG), bent into a loop at the tip and mounted in a suitable handle, makes a good inoculating loop. Immediately before (and after) use it should be heated to red heat and allowed to cool.

Bottle caps and necks should be flamed as they are opened. The lids of petri dishes should be lifted only sufficiently far to allow transfer using the loop-tip of the inoculating loop.

Care should be taken when sterilising the loop after use to make sure that the medium and organisms are not splattered by rapid heating as this can rapidly contaminate a laboratory or transfer cabinet.

Maintenance of cultures
It may be most convenient to maintain cultures for short periods of time only. Keeping them at a low temperature reduces the growth rate and many, if kept in a refrigerator, will survive in a viable condition for several months. Sub-culturing will be necessary at intervals.

Aspergillus
Maintain as slope cultures on nutrient agar (see page 109).

Bacillus subtilis
Maintain on nutrient agar (see page 109) or in nutrient broth (see page 110).

Coprinus
The cultures will normally be received as vegetative slopes of the haploid strains growing on complete medium. They can be kept in this form for up to six months in a refrigerator without sub-culture. Adenine requiring strains, however, need sub-culturing more frequently.

The culture media are probably best obtained commercially ready for use where this is possible but the recipes are given in Appendix 3 for the benefit of those who prefer to make up their own.[104]

To get fruiting bodies from dikaryons on sterilised horse dung
Collect dung and dry it by spreading it out on a newspaper in a warm dry place, for example a boiler room or greenhouse. The dung can be kept indefinitely in a dry state.

Add a little dry dung to about 40cm^3 of water in a glass container, e.g. a jam jar; cover with metal foil and sterilise in an autoclave.

Inoculate with dikaryon mycelium and incubate at 37°C for one day to give growth a start.

Place in a light, warm place, about 28°C if possible. Fruiting bodies begin to appear in about 7-14 days.

For demonstration purposes it is best to make a series of inoculations at 2-3 day intervals.

Fusarium
Maintain as slope cultures on malt agar (see page 109).

Saccharomyces
Maintain as slope cultures on complete medium (see page 108).

Sarcina
Maintain on nutrient agar (see page 109).

Sordaria
Maintain as slope cultures on malt or cornmeal agar. Keep in a refrigerator when sufficient growth has been made.

Appendix 2 Chemical recipes

The solutions, media and nutrient mixtures which are available ready-made from biological suppliers or other sources are indicated by an asterisk.

**Acetic alcohol* (for chromosome preparations)
25cm^3 acetic acid (glacial)
75cm^3 ethanol or industrial spirit

**Acid-alcohol* (for pigment extraction in *Antirrhinum*)
1cm^3 concentrated hydrochloric acid
99cm^3 ethanol or industrial spirit

Buffer solutions

1 Buffer/formalin
 2.0g formalin (approximately 2.0cm^3 40% solution solution of formaldehyde in water)
 100cm^3 phosphate buffer 0.067M
2 Buffer/phosphate 0.067M pH 7.3.
 0.75g Disodium hydrogen phosphate (anhydrous)
 0.18g Potassium dihydrogen phosphate (anhydrous)
 100cm^3 distilled water

* *Composts*
John Innes composts
These are prepared from the following:
A Medium sieved loam partially sterilized to remove harmful organisms. This can be achieved by heating at between 50 and 100°C for 10 minutes.
B Sieved granular or fibrous peat lightly moistened. (Both loam and peat should be passed through a 1 cm sieve.)
C Dry coarse sand.
D Superphosphate of lime.
E Ground limestone or chalk.
F John Innes base made by mixing well the following parts by weight:
 2 parts hoof and horn meal (13 per cent N)
 2 parts superphosphate of lime (18 per cent P_2O_5)
 1 part sulphate of potash (48 per cent K_2O)

Type of compost	parts by volume			Kg/m^3 of the mixture		
	A	B	C	D	E	F
seed	2	1	1	0.9	0.45	
potting No 1	7	3	2		0.45	2.4
potting No 2	7	3	2		0.9	4.7
potting No 3	7	3	2		1.35	7.0

Preparing the compost
The loam (A) should be spread out on a clean dry floor with the peat (B) and sand (C) on top. The fertilizers (D, E, F) are sprinkled over the surface to ensure even distribution and the heap turned over three or four times.

Use
Potting compost No 2 is used for plants growing in large pots and No 3 for plants of a particularly vigorous nature.

Chromatogram solvents

1 Butanol/acetic acid solvent
 60 cm^3 n-butanol (butan-1-ol)
 10 cm^3 glacial acetic acid
 20 cm^3 distilled water
2 Butanol/ammonia solvent
 172 cm^3 n-butanol (butan-1-ol)
 18 cm^3 distilled water
 10 cm^3 ammonium hydroxide solution (0.880 ammonia)
3 Isopropanol/hydrochloric acid solvent
 170 cm^3 isopropanol (propan-2-ol)
 39 cm^3 distilled water
 41 cm^3 concentrated hydrochloric acid

Ethylmethane sulphonate. EMS
1 cm^3 ethylmethane sulphonate
120 cm^3 distilled water

This gives a solution of approximately 1% concentration; lower concentrations can be prepared by dilution. Prepare immediately before use. The solution is stable for about a week if kept in a refrigerator. Note that EMS is a carcinogen and must be treated with the strictest precautions.

Foliar sprays for plants

1 Pyrimidine foliar spray for tomatoes.
 0.1 g pyrimidine
 1.0 dm^3 distilled water.
 Keep solution in refrigerator.

2 Thiamine (Vitamin B_1) foliar spray for tomatoes.
 0.1 g thiamine
 1.0 dm^3 distilled water.
 The solution may be kept for up to two weeks in a refrigerator, though it is preferable to make a fresh solution for each use.

* *Glycerine-albumen*

50cm^3	glycerol
1g	sodium salicylate
50cm^3	white of hen's egg

Separate the white from the yolk of the egg, dissolve the salicylate in 0.5–1.0 cm^3 of water, and add the other ingredients. Mix well and filter before use.

Growth regulating solutions

1 Gibberellic acid solution (10 ppm)
 0.01 g gibberellic acid
 1.0 dm^3 distilled water
 Dissolve the gibberellic acid in 2 cm^3 ethanol and make up to 1 dm^3 with distilled water. Store the solution in a refrigerator.

2 Indolylacetic acid
 0.01 g indolylacetic acid (indol-3-yl-acetic acid)
 1.0 dm^3 distilled water
 Dissolve the indolylacetic acid in 2.0 cm^3 ethanol and add to 900 cm^3 warm distilled water at 80°C. Finally make up to 1 dm^3. Store the solution in a refrigerator.

Humidity regulating solutions

1 *Sulphuric acid solutions*

Relative humidity % at 25°C	*Weight of sulphuric acid in g per 100 g of water*
100	0
95	11.0
90	17.9
85	22.9
80	26.8
75	30.1
70	33.1
65	35.8
60	38.4
55	40.8
50	43.1
45	45.4
40	47.7
35	50.0
30	52.5
25	55.0
20	57.7
15	60.8
10	64.5
5	69.4

2 *Saturated salt solutions*

Relative humidity %	*Salt solution*
98.0	potassium dichromate
92.5	potassium nitrate
90.2	barium chloride
84.3	potassium chloride
80.7	potassium bromide
75.3	sodium chloride
73.8	sodium nitrate
70.8	strontium chloride
57.7	sodium bromide
52.9	magnesium nitrate
47.1	lithium nitrate
42.8	potassium carbonate
33.0	magnesium chloride
22.5	potassium acetate
11.1	lithium chloride
7.0	sodium hydroxide

3 *Glycerol and water mixtures*

Relative humidity%	*% glycerol/water (weight/weight)*
90	33
80	51
70	64
60	72
50	79
40	84
30	89
20	92
10	95

* *Hydrochloric acid: molar (M)*

86.0 cm^3	concentrated hydrochloric acid
914.0 cm^3	distilled water

tenth molar (0.1M)

8.6 cm^3	concentrated hydrochloric acid
991.4 cm^3	distilled water

Exercise extreme caution when mixing.

Media

1 *Agar media*

A solid medium may be prepared by dissolving 10-15 g of agar in 1 dm^3 of distilled water.

Slightly more agar is required when preparing media with acidic additives. 15-20 g/dm^3 will generally be found to be satisfactory.

When the agar is dissolved the medium is autoclaved to 121°C (103.5kN/m^2≏15lbf/in^2) for fifteen minutes. It may be necessary to filter before autoclaving.

The autoclaved medium is allowed to cool to 45 to 50°C before pouring into Petri dishes or test tubes for slope cultures.

1.1 Complete medium for *Saccharomyces*

15.0 g	agar
20.0 g	glucose
100.0 cm^3	stock solution 1 inorganic salts (see page 112)
1.0 cm^3	stock solution 2 calcium chloride (see page 112)
1.0 cm^3	stock solution 3 vitamins (see page 112)
0.000 02 g	biotin
0.01 g	uracil
0.01 g	adenine

Make up to 1 dm^3 with distilled water.

Melt 15 g agar in about 500 cm^3 of distilled water in a pan. Mix 20 g glucose and stock solutions 1, 2, and 3 in amounts indicated, with 300 cm^3 of distilled water in another container. Pour melted agar into it, stir and add distilled water to make 1 dm^3. Pour media into flasks when hot and sterilize within two hours.

Use 20 cm^3 of media per Petri dish. Thus 50 plates can be obtained from 1 dm^3.

The plates can be stored indefinitely in a refrigerator. If the lids are sealed with Sellotape, they can be stored in an ordinary cupboard for about a week—perhaps more—without fear of contamination.

Unless a large number of plates is to be prepared it is most economical in time and money to purchase the culture medium already made up in sterile Petri dishes. These can be obtained from biological supply agencies.

1.2 *Cornmeal agar for *Sordaria*

30 g	maize meal
1 dm^3	distilled water

Boil for 15 minutes and then decant off the clear liquid. Add 2.0 g agar to each 100 cm^3 of clear liquid.

or

17 g	Difco cornmeal agar (obtainable from Oxoid Ltd)
1 g	yeast extract
1 dm^3	distilled water

1.3 *Dox's Agar for *Fusarium*

0.01 g	ferrous sulphate
0.05 g	magnesium sulphate
0.05 g	potassium chloride
1.0 g	potassium dihydrogen phosphate
2.0 g	sodium nitrate
15.0 g	sucrose
15.0 g	agar
1 dm^3	distilled water

1.4 Glucose-1-phosphate agar

2.0 g	agar
0.5 g	glucose :1: phosphate
100 cm^3	distilled water

Mix ingredients together thoroughly. Pour plates and heat gently to about 50°C in Petri dishes to a depth of no more than 0.2 cm. Sufficient for about 15 dishes.

1.5 *Magnesium deficient agar for *Aspergillus*
As for sporulation agar but without the magnesium sulphate.

1.6 *Malt agar

15 g	agar
20 g	malt extract
1 dm^3	distilled water

1.7 *Minimal agar medium (Fries) for *Coprinus* (see also page 110)

20 g	glucose
2 g	asparagine
25 cm^3	stock salt solution (see page 110)
4 cm^3	thiamine (0.001g/100cm^3)
20 g	agar
1 dm^3	distilled water

Weigh out asparagine and dissolve in 100 cm^3 distilled water using only very gentle heat.
Dissolve glucose in 200 cm^3 distilled water.
Weigh out thiamine 0.005g and dissolve in 500 cm^3 distilled water, take 4 cm^3 of solution.
Add these three to 25 cm^3 of the stock salt solution and make up to 1000 cm^3.
Add agar and stir in open autoclave until agar is broken down. Dispense into medical flats.
Sterilise in autoclave at 120°C for 15 minutes.

1.8 Nagai medium for *Saccharomyces*

20.0g	glucose
1.5 g	peptone
1.5 g	yeast extract
1.5 g	potassium dihydrogen phosphate (KH_2PO_4)
1.0 g	magnesium sulphate ($MgSO_4$)
1.5 g	ammonium sulphate $(NH_4)_2SO_4$
12.0 g	agar

Dissolve in 1 dm^3 of distilled water.
Heat in water bath till the agar has melted; then autoclave at 121°C (103.5kN/m^2 ≙ 15 lbf/in^2) for 15 minutes.

1.9 Magdala red medium for *Saccharomyces*
Magdala red stock solution
Prepare by dissolving Magdala red in distilled water at the rate of 0.001 g per cm^3.
When required sterilize 10 cm^3 of the stock solution in a water bath at 100°C for 1 hour.
After autoclaving allow 1 dm^3 Nagai medium to cool to approximately 55°C and then add 10 cm^3 Magdala red solution aseptically. Mix. Pour plates which have a dye concentration of 10 ppm at a pH around 5.4.

1.10 *Nutrient agar for complete growth of *Aspergillus*

20 g	malt extract
10 g	peptone
20 g	glucose
15 g	agar
1 dm^3	distilled water.

1.11 Nutrient agar for *Bacillus subtilis* and *Sarcinia lutea*

15 g	agar
1 dm^3	nutrient broth

1.12 *Potato dextrose agar for *Fusarium*

250 g	potatoes
20 g	glucose
15 g	agar
1 dm^3	tap water

Wash potatoes thoroughly and dice but do not peel them. Place in muslin bag and steam for one hour. Suspend bag over water and allow liquid to drip from it without squeezing. Make up extract to 1 dm^3 add agar and glucose then autoclave.

1.13 *Sporulation agar for *Aspergillus*

1.0 g	sodium nitrate ($NaNo_3$)
0.5 g	magnesium sulphate ($MgSO_4.7H_20$)
0.5 g	potassium chloride (KC1)
1.5 g	potassium dihydrogen phosphate (KH_2PO_4)
trace	iron (II) sulphate ($FeSO_4.H_2O$)
trace	zinc sulphate ($ZnSO_4$)
20.0 g	glucose
15.0 g	agar
1 dm^3	distilled water

1.14 Thioglycollate agar for *Bacillus subtilis* and *Sarcina lutea*

15 g	agar
1 dm^3	thioglycollate broth

2 *Broths*

Broths are liquid media for the culture of bacteria and some fungi. Heat to 65°C to dissolve ingredients and filter. Autoclave to 121°C (103.5kN/m^2≏15lbf/in^2 for fifteen minutes before pouring into suitable containers. Allow to cool before inoculating.

2.1 *Nutrient broth for *Bacillus subtilis* and *Sarcina lutea*

10 g	beef extract
10 g	peptone
5 g	sodium chloride
1 dm^3	distilled water

Adjust pH to 7.4 with sodium bicarbonate.

2.2 Thioglycollate broth for *Bacillus subtilis* and *Sarcina lutea*

1.0 g	sodium thioglycollate
10.0 g	glucose
0.5 g	powdered agar
0.2cm^3	methylene blue
1 dm^3	distilled water

Adjust pH to 7.4 with sodium bicarbonate.

Nitrogenous base solutions

1 Adenine, cytosine, guanine or thymine

1 mg	base
5 cm^3	0.1M hydrochloric acid

2 Deoxyribonucleic acid solution

2 mg	DNA
1 cm^3	distilled water

Nutrient mixtures for individual organisms

See also media.

1 *Coprinus lagopus*

1.1 *Stock salt solution

10.0 g	ammonium tartrate
20.0 g	potassium dihydrogen phosphate
5.8 g	sodium sulphate
45.0 g	sodium hydrogen phosphate

Make up to 500 cm^3 with distilled water.

1.2 *Hydrolysed nucleic acid

1 g	yeast nucleic acid	in 15 cm^3
1 g	thymus nucleic acid	N. NaOH
1 g	yeast nucleic acid	in 15 cm^3
1 g	thymus nucleic acid	N. HCL

Place the two samples in small bottles without caps. Autoclave to 107°C (34.45kN/m^2≏5lbf/in^2) for 20 minutes. Mix together. Adjust pH to 6.0. Filter hot. Make up to 40 cm^3 with distilled water.
Store under chloroform in dark bottle.

1.3 Complete medium (Fries)

To 1000 cm^3 of minimal medium add

0.75 g	yeast extract
0.75 g	eupepton
0.60 g	malt extract
1.25 cm^3	hydrolysed nucleic acid solution (1.2)

1.4 Supplements to minimal medium for the appropriate mutants

Mutant	Supplement(s)	g or cm^3/100 cm^3 stock solution	cm^3 stock solution/1000 cm^3 medium
ad 8	adenine	0.02 g	50
paba 1 & paba 2	para-aminobenzoic acid	0.005 g	10
met 9	methionine	0.10 g	10
chol 1 & chol 2	choline chloride	0.02 g	10
	dimethyl ethanolamine	0.10 cm^3	10
	monomethyl ethanolamine	0.10 cm^3	10
	ethanolamine	0.10 cm^3	10
nic 4	nicotinic acid	0.02 g	10

2 *Drosophila*

The food base must pour easily when hot and set and remain firm when cold. Slight modifications to the medium may be necessary according to the quality of the ingredients.

Medium 1

Soak 72 g of oatmeal in 120 cm^3 of water
Dissolve 35 g of black treacle in 40 cm^3 of water
Boil 6 g of agar in 400 cm^3 of water

Mix all these together and add a pinch of Nipagin (methylhydroxybenzoate). Boil with constant stirring for at least 15 minutes. If the hot mixture does not pour easily add more water and continue stirring until the right consistency is reached.
Pour base into sterilised bottles or tubes, 2.5 cm deep in bottles and 1.5 cm in tubes. It is best to avoid medium getting on the sides of the containers. This can be facilitated by using a wide glass tube drawn to a blunt point and fitted with a rubber bulb as a giant pipette, especially if the pipette is kept warm by wrapping it in a hot cloth. Plug containers immediately medium is poured in.

When the medium is set, 'seed' the surface with a few drops of a thick suspension of baker's yeast in water. This quantity is sufficient for about ten small milk bottles or fifty specimen tubes.

Medium 2

agar	9 g
powdered dried yeast	16 g
sucrose	27 g
proprionic acid	2.5 cm^3
water	500 cm^3

Boil with constant stirring the agar, yeast and sucrose with the water till dissolved then add the proprionic acid. 'Seed' the surface as for medium 1.

3 *Locusts*

3.1 Protein supplement. Dry wheat bran and powdered dried yeast in proportions of about 5:1. Thin flaked maize may also be given. Prepared high protein foods such as 'Bemax' are also suitable.

3.2 Artificial dry diet assuming locusts have free access to water
1 part by volume powdered dried yeast
10 parts by volume dried milk
10 parts by volume dry wheat bran
10 parts by volume dried grass

3.3 Fresh food is the best. Use growing barley, wheat, oats or a quick growing grass cultivar such as *Lolium multiflorum* 'Westernwolth'.

4 *Tribolium*

1 part by weight powdered dried yeast
10–12 parts by weight wholemeal or wheatmeal flour.

It is best to heat sterilise to avoid mite contamination. A drinking fountain may be added to the stock culture container to increase humidity and give a supply of water.

5 *Saccharomyces cerevisiae*

*Stock solutions for complete medium

5.1 Stock solution 1—inorganic salts

10.0 g	ammonium sulphate ($(NH_4)_2SO_4$)
8.75 g	potassium hydrogen phosphate (KH_2PO_4)
1.25 g	dipotassium dihydrogen phosphate K_2HPO_4
5.0 g	magnesium sulphate ($MgSO_4.7H_2O$)
1.0 g	sodium chloride (NaCl)

0.1 cm^3 of 0.1 per cent stock solution boric acid (H_3BO_3)
0.1 cm^3 of 0.1 per cent stock solution copper sulphate ($CuSO_4.5H_2O$)
0.1 cm^3 of 0.1 per cent stock solution potassium iodide (KI)
0.1 cm^3 of 0.5 per cent stock solution iron (III) chloride ($FeCl_3.6H_2O$)
0.1 cm^3 of 0.7 per cent stock solution zinc sulphate ($ZnSO_4.7H_2O$)
Dissolve the above in distilled water to make 1 dm^3 of stock solution 1.

5.2 Stock solution 2—calcium chloride. Dissolve 10 g calcium chloride ($CaCl_2.2H_2O$) in distilled water to make 100 cm^3 of stock solution 2.

5.3 Stock solution 3—vitamins

0.04 g	thiamin hydrochloride (aneurine)
0.04 g	pyridoxine hydrochloride
0.2 g	inositol
0.04 g	D-pantothenic acid calcium salt

Dissolve the above in distilled water to make 100 cm^3 of stock solution 3.
Where very small amounts of chemicals are required it is possible to use an estimated amount and obtain reasonable results. This saves the tedium of delicate weighing, or helps when a delicate balance is not available.

Pre-treatment solutions for root tips

1 Paradichlorbenzene

5 g	paradichlorbenzene
500 cm^3	distilled water

Stand overnight at about 60°C and filter before use.

2 Colchicine

0.2 g	colchicine
100 cm^3	distilled water

Physiological saline solution (for preparation of the salivary glands of *Drosophila* larvae)
0.67% aqueous sodium chloride

Stain solutions

1 *Acetic-carmine

45 cm^3	acetic acid (glacial)
1 g	carmine
55 cm^3	distilled water

Agitate the carmine in the acetic acid, add the water, and bring to the boil. Cool and filter off excess carmine. (This stain may be used with or without a mordant. If a specimen in the stain is teased with iron needles, sufficient iron dissolves from the needles to act as a mordant. A drop or two of 45% acetic acid saturated with iron acetate can be added as the mordant).

2 *Acetic-lacmoid

The stain is an indicator and is known as resorcin blue as a dye.
Prepare as for acetic-orcein.

3 *Acetic-orcein

The stain orcein deteriorates in dilute acid and so it is best to prepare it from dry stain or keep it as a concentrated solution which must be diluted for use. A greater concentration is required with synthetic orceins.

100 cm^3	acetic acid (glacial)
2.2 g	orcein

Dissolve the orcein in the acid by gently boiling for about 6 hours, using a reflux condenser. Filter and bottle to form a stock solution.

To use, mix:

9 cm^3	stock solution
11 cm^3	water

Only dilute sufficient for immediate use.

4 *Aniline blue

1 g	aniline blue water soluble.
99 cm^3	distilled water (or 70% ethanol).

5 *Feulgen stain (Schiff's reagent)

200 cm^3	distilled water
1 g	fuchsin basic
30 cm^3	M hydrochloric acid
3 g	potassium metabisulphite ($K_2S_2O_5$)

Preparation Bring distilled water to the boil and add the fuchsin. Shake well and cool to 50°C. Add the M hydrochloric acid and the potassium metabisulphite. Allow to bleach for 24 hours in a tightly stoppered bottle in the dark. Add 0.5 g decolorizing charcoal. Shake thoroughly and filter rapidly through coarse filter paper. Store in a tightly stoppered bottle in a cool, dark place.

6 Iodine solution

1 g	iodine crystals
6 g	potassium iodide

Dissolve the potassium iodide in 200 cm^3 distilled water and then add the iodine crystals. When they are dissolved make up to 1 dm^3 with distilled water.

7 Methyl green pyronin in acetate

0.545 g	sodium acetate
10.0 cm^3	glacial acetic acid
1.0 g	methyl green pyronin
	distilled water

Dissolve the sodium acetate in the glacial acetic acid. Add to 85 cm^3 distilled water and dissolve the methyl green pyronin. Finally make up to 100 cm^3 with distilled water.

Appendix 3 Addresses of sources of information and of suppliers

An extensive listing will be found in reference 10.

3.1 Information
Publication, societies and other sources

British Rabbit Council,
Purefoy House,
7 Kirkgate,
Newark,
Nottinghamshire NG24 1AD

Cage & Aviary Birds Diary (Yearly)
Published by Castell Bros Ltd.,
15 St Cross Street,
London EC1N 8UT
Lists all the specialist societies for agriculture

Catalogue of the Chelsea Flower Show (Yearly)
Royal Horticultural Society,
Horticultural Hall,
Vincent Square,
London SW1P 2PE
Contains many addresses of suppliers of plants and horticultural equipment

Centre for Overseas Pest Research,
College House,
Wrights Lane,
London W8 5SJ
(Formerly the Anti-Locust Research Centre). Leaflet on locust allergy

Fur and Feather
High Street,
Idle,
Bradford,
Yorkshire BD10 8NL
A list of the various specialist breed clubs, both national and regional, for small mammals is given at intervals.

The Genetical Society,
Secretary: Professor D.A. Hopwood,
John Innes Institute,
Colney Lane, Colney,
Norwich NOR 7OF

Mongolian Gerbil Society,
Secretary: K.W. Smith,
6 Duchy Close,
Higham Ferrers,
Northamptonshire NN9 8BZ

M R C Laboratory Animals Centre,
Medical Research Council Laboratories,
Woodmansterne Road,
Carshalton,
Surrey SM5 4EF

National Cavy Club,
Secretary: D. Parkinson,
23 Union Street,
Slaithwaite,
Huddersfield,
Yorkshire HD7 5ED

National Hamster Council,
Secretary: P.W. Parslow,
Parslow's Hamster Farm,
Commonside,
Great Bookham,
Nr Leatherhead,
Surrey KT23 3JZ

National Mouse Club,
Secretary: L. Heywood,
31 Bradbury Street,
Barnsley,
Yorkshire S70 6AQ

The National Sweet Pea Society
Secretary: R.J. Huntley,
33 Priory Road,
Rustington,
Sussex BN16 3PZ

Poultry Club,
Secretary: Mrs S. Jones,
72 Springfields,
Dunmow,
Essex CM6 1BS

Yearbook lists affiliated clubs and societies and breeders

Poultry World
161/166 Fleet Street,
London EC4 4AA
International directory of suppliers of stock and associated equipment
published yearly in August

The Royal National Rose Society,
Chiswell Green Lane,
St Albans,
Hertfordshire AL2 3NR
Publications include the Rose Annual, Rose Bulletin and those on cultivation and lists of varieties

Royal Horticultural Society Gardener's Diary and Notebook (Yearly)
Published by Charles Letts & Co Ltd,
Diary House,
Borough Road,
London SW1 1DW

Lists all Government departments and some associations concerned with horticulture and preservation; botanic gardens; horticultural research stations and specialist horticultural societies

3.2 **General biological supplies**
These firms supply a wide range of living organisms suitable for genetic use, associated apparatus and equipment.

Bioserv Ltd.,
38 Station Road,
Worthing,
Sussex, BN11 1JP

Carolina Biological Supply Company,
Elon College,
Burlington,
N. Carolina 27215,
USA
(UK Agents T. Gerrard & Co.)
Publications available from the UK Agents include *Carolina Tips*

GBI (Labs) Ltd.,
Heaton Street,
Denton,
Manchester,
Lancashire M34 3RG

T. Gerrard & Co.
(Incorporating Griffin Biological Laboratories)
Gerrard House.
Worthing Road,
East Preston,
West Sussex BN16 1AS

Philip Harris Biological Ltd,
Oldmixon,
Weston-super-Mare,
Somerset BS24 9BJ

Timstar Biological Suppliers,
Lower House,
Little Budworth,
Tarporley,
Cheshire CW6 9BL
Particularly living organisms

Turtox/Cambosco/,
Macmillan Science Incorporated,
8200 South Hoyne Avenue,
Chicago,
Illinois,
USA
(UK Agent Griffin Biological Laboratories)

Publications available from UK agents include 'Turtox News' and a series of Service Leaflets

Wards Natural Science Establishment Incorporated,
PO Box 1712,
Rochester 3,
New York,
USA
(UK Agents Philip Harris Biological Ltd)
Leaflets on culture techniques available from UK Agents

3.3 Cages, racking and shelving units, hutches and aquarium frames

See also General Biological Suppliers 3.2

All Type Tools (Woolwich) Ltd,
Animal Cages Dept,
Purland Road,
London SE28 0AF
Metal and plastic mammal cages

Aluminium Equipment Co Ltd.,
21a Conewood Street,
Highbury,
London N5 1BZ
Locust cages

Associated Crates Ltd,
Holloway Bank,
Wednesbury,
Staffs WS10 0NL
Metal and plastic mammal cages

Bowman, E.K. Ltd,
32, 34 and 57 Holmes Road,
London NW5 3AG
Metal and plastic mammal cages

Cope & Cope Ltd,
57 Vastern Road,
Reading,
Berkshire RG1 8BX
Metal and plastic mammal cages

North Kent Plastic Cages Ltd,
Home Gardens,
Dartford,
Kent DA1 1EQ
Metal and plastic mammal cages

Stoners Timber Buildings Ltd,
Fleming Way,
Crawley,
Sussex RH10 2LH
Aviaries

Sydenham Hannaford, C.A. Ltd,
Hamworthy Junction,
Poole,
Dorset BH16 5BP
Rabbit hutches and accessories; poultry houses

3.4 Chemicals, culture media and stains

Astell Laboratory Service Co., (Astell-Hearson)
172 Brownhill Road,
Catford,
London SE6 2DL
Culture media

Cambrian Chemicals Ltd,
Suffolk House,
George Street,
Croydon,
Surrey CR9 3QL
Biochemicals, including enzymes and ribonucleosides

Difco Laboratories,
P.O. Box 14B,
Central Avenue,
Molesey,
Surrey KT8 0SE
Culture media and reagents; biological stains

Flow Laboratories Ltd,
Victoria Park,
Heatherhouse Road,
Irvine,
Ayrshire KA12 8NB
Tissue cultures, tissue culture media and sera, viral and immunological reagents; microtitration equipment, tissue culture plastics and laboratory equipment

Hopkins and Williams Ltd,
Freshwater Road,
Chadwell Heath,
Essex RM1 1HA
Laboratory chemicals

Horwell, Arnold R,
2 Grangeway,
Kilburn High Road,
London NW6 2BP
Blood grouping sera

Hughes & Hughes Ltd,
Elms Industrial Estate,
Church Road,
Harold Wood,
Romford,
Essex RM3 OHR
Biochemicals including enzymes; blood grouping and typing sera

Koch Light Laboratories Ltd,
Colnbrook,
Buckinghamshire SL3 0BZ
Organic chemicals including ATP, enzymes and indicators and lesser used inorganic chemicals

Lamb, Raymond A.,
6 Sunbeam Road,
London NW10 6JL
Stains, chemicals and waxes for microscopy; microscope accessories

May & Baker Ltd,
Dagenham,
Essex RM10 7XS
Laboratory chemicals

Microbiological Supplies,
P.O. Box 10,
Tunbridge Wells,
Kent TN1 1SZ
Culture media

Northern Media Supply Ltd,
Crosslands Lane,
Newport Road,
North Cave,
Brough,
E. Yorkshire HU15 2PG
Media

Oxoid Ltd,
Wade Road,
Basingstoke,
Hampshire RG24 0PW
Culture media, Ringers solution tablets, discs impregnated with antibiotics, accessories for microbiology

Searle Scientific Services, (G.T. Gurr)
Coronation Road,
Cressex Industrial Estate,
High Wycombe,
Buckinghamshire HP12 3TA
Biological stains

Sigma London, Chemicals Co. Ltd,
Norbiton Station Yard,
Kingston upon Thames,
Surrey KT2 7BH
Amino acids, enzymes, hormones, vitamins

3.5 Specialist equipment

Camlab (Ltd),
Nuffield Road,
Cambridge CB4 1TH
Chromatographic apparatus, ovens

Grundy Equipment Ltd,
Packet Boat Lane,
Cowley Peachey,
Uxbridge,
Middlesex UB8 2JL
Incubators, ovens and waterbaths

Halsey's Electrical Wholesale Co. Ltd,
Brandon House,
Wyfold Road,
London SW6 6SQ
Ultra-violet lamps

Hanovia Ltd,
480 Bath Road,
Slough,
Buckinghamshire SL1 6BL
Ultra-violet lamps

Hearson, Charles & Co., (Astell-Hearson),
172 Brownhill Road,
Catford,
London SE6 2DL
Ovens, incubators, waterbaths

Northern Media Supply Ltd,
Crosslands Lane,
Newport Road,
North Cave,
Brough,
E. Yorkshire HU15 2PG
Autoclaves, incubators and waterbaths.

Pickstone, R.E. Ltd,
Faraday Place,
Thetford,
Norfolk
Incubators, ovens, sterilisers

Shandon Southern Products Ltd,
93/96 Chadwick,
Astmoor Industrial Estate,
Runcorn,
Cheshire WA7 1PR
Chromatography and other equipment

Sydenham Hannaford, C.A. Ltd,
Hamworthy Junction,
Poole,
Dorset BH16 5BB
Incubators and brooders

Western Incubators Ltd,
Springfield Industrial Estate,
Springfield Road,
Burnham-on-Crouch,
Essex CM0 8TA
Western-Curfew incubators

3.6 Living organisms

The General Biological Suppliers (3.2) have a wide range of living organisms available, only those specialist suppliers are listed here.

When ordering living organisms adequate prior notice of requirements must be given. This is especially so for those only available in season, for example, insects and certain plants.

3.6.1 *Mammals*

Small mammals are best obtained from accredited breeders or recognised suppliers registered with the MRC Laboratory Animals Centre. A list of these may be obtained from the Centre. Farm animals should be obtained only from reputable commercial suppliers.

MRC Laboratory Animals Centre,
Medical Research Council Laboratories,
Woodmansterne Road,
Carshalton,
Surrey SM5 4EF

3.6.2 *Tropical Fish*

Griffin Biological Laboratories.,
Gerrard House,
Worthing Road,
East Preston,
West Sussex BN16 1AS
Genetic strains of *Lebistes maculatus*

3.6.3 *Birds*

Avon Aviaries,
34 Port Street,
Evesham,
Worcestershire WR11 6AW

Bleak Hall Bird Farm (Luton) Ltd,
12, Cresta House,
Alma Street,
Luton,
Bedfordshire LU1 2PL

Fylde Foreign Bird Farm,
442 Midgeland Road,
Marton Moss,
Blackpool,
Lancashire FY4 5EF

Keston Foreign Bird Farm Ltd.,
Brambletye,
Keston,
Kent BR2 6AQ

Ponderosa Bird Aviaries,
The White House,
Branch Lane,
The Reddings,
Cheltenham,
Gloucestershire GL51 6RP
Associated equipment and foods

Severndown Bird Farm,
The Cottage,
10 Goose Green,
Yate,
Bristol BS17 5BJ

Southern Aviaries Ltd,
Brook House Farm,
Tinkers Lane,
Hadlow Down,
Nr Uckfield,
Sussex TN22 4EU

Taylor, D.J.
Newmarket Cottage,
Clay Lane,
Clay Cross,
Chesterfield,
Derbyshire S45 9AP

Toddington Bird Farm,
11 Station Road,
Toddington,
Dunstable,
Bedfordshire LU5 6BN

The advertisment section in *Poultry World* (3.1) should be consulted for suppliers of fertile hen's eggs and poultry.

3.6.4 *Insects*
Larujon Locust Suppliers,
c/o Welsh Mountain Zoo,
Colwyn Bay,
North Wales LL28 5UY
Locusta migratoria and *Schistocera gregaria*

3.6.5 *Flowering plants and ferns*
A range of plants may usually be obtained from local horitcultural sundriesmen and nurseries. Specialist suppliers only are listed.

Bees Ltd,
Sealand,
Chester,
Cheshire CH1 6BA
Herbaceous plants and roses

Blom, Walter & Son,
Coombelands Nurseries,
Leavesden,
Watford,
Hertfordshire WD2 7BH
Bulbs and corms (seeds)

Clifton Geranium Nurseries,
Earnley Gardens Ltd.,
Cherry Orchard Road,
Chichester,
Sussex PO19 2BX
Cultivars of *Geranium* and *Pelargonium*

Gamble D. & Sons,
Highfield Nurseries,
Longford,
Derby DE6 3DT
Cultivars of *Pelargonium*

Gregory C. & Sons Ltd,
The Rose Garden,
Stapleford,
Nottingham NG9 7JA
Cultivars, including miniatures, of *Rosa*

Greybridge Geraniums,
Fibrex Nurseries Ltd.,
Harvey Road,
Evesham,
Worcestershire WR11 5AQ
Cultivars and species of *Geranium* and *Pelargonium*

Hillier and Sons,
Winchester,
Hampshire SO22 5DN
Herbaceous plants, trees and shrubs

Pedley, S. & Sons,
Preston Road Nursery,
Newton,
Preston,
Lancashire PR4 3RL
Cultivars of *Coleus* and *Hedera*

Price, Mary E.,
Fernhurst,
Roncarbery,
Co. Cork,
Ireland
Asplenium adiantum-nigrum, A. trichomanes, Blechnum spicant, Cterach officinarum, Osmunda regalis, Phyllitis scolopendrium, Polypodium vulgare

Russell, L.R. Ltd,
Richmond Nurseries,
Windlesham,
Surrey GU20 6LL
Shrubs, trees, dwarf hardy perennials

Waterer, John, Sons and Crisp Ltd,
The Floral Mile,
Twyford,
Berkshire RS10 9SJ
Herbaceous plants and roses

Welsh Plant Breeding Station,
Plas Gogerddan,
Nr. Aberystwyth,
Wales SY23 3EB
Trifolium repens with multiple allele leaf markings

3.6.6 *Seeds and fern spores*
The General Biological Suppliers (3.2) keep a range of seeds suitable for genetic investigations and also irradiated seed.

Seeds from a number of other firms, in addition to some of those listed, are available from local horticultural sundriesmen.

Burlingham, George & Sons Ltd,
Station Hill,
Bury St Edmunds,
Suffolk IP32 6AE
Clover, cereals, herbage and some root crops

Carters Tested Seeds Ltd,
Lower Dee Mills,
Llangollen,
Denbighshire LL20 8SD

Dobie, Samuel & Sons Ltd,
Upper Dee Mills,
Llangollen,
Denbighshire LL20 8SD
Mimosa pudica, also fern spores

Elsoms (Spalding) Ltd,
Elsom House,
Spalding,
Lincolnshire PE11 1TD
Clover, cereals, herbage and vegetables

Howell, Major V.F.,
'Firethorn',
6 Oxshott Way,
Cobham,
Surrey KT11 2RT
Wide range of the more rare seeds

Practical Plant Genetics,
18 Harsfold Road,
Rustington,
Sussex BN16 2QE
Tomato seed for genetics

Sutton & Sons Ltd,
Royal Seed Establishment,
London Road,
Reading,
Berkshire RG6 1AB

Thompson and Morgan of Ipswich Ltd,
London Road,
Ipswich,
Suffolk IP2 0BA

Unwin, W.J. Ltd,
Histon,
Cambridgeshire CB4 4LE
Also range of plants

3.6.7 *Micro-organisms: fungi and bacteria*
Commonwealth Mycological Institute,
Collection of Fungus Cultures,
Ferry Lane,
Kew,
Surrey TW9 3AF
Fungi (other than animal pathogens, wood rotting species and most yeasts). A leaflet listing those fungi suitable for teaching purposes is available.

Microbiological Supplies,
P.O. Box 10,
Tunbridge Wells,
Kent TN1 1SZ

Appendix 4 Bibliography and references

Section 1 General and Apparatus 1–26
Including books describing the genetic use of a variety of organisms.

1 Beaumont, B.S., Ellis, A & Bryant, J.J. (Eds) (1964). *The Science Masters Book, Series 4 Part 3, Biology*. John Murray.
2 Biological Sciences Curriculum Study. (1965). *Biological Science: Interaction of Experiments and Ideas.* Students Text (also Teachers Guide). Prentice-Hall.
3 Bradbury, S. (1973). *Peacock's Elementary Microtechnique.* 4th edn. E. Arnold.
4 Brierley, J.K. (1961). Some Suggestions for the Teaching of evolution in the field, garden & laboratory. *School Science Review* 42 (148), 401-10.
5 Clarke, R.A., Booth, P.R., Grigsby, P.E., Haddow, J.F & Irvine, J.S. (1968). *Biology by Inquiry*, Book 1 Students Text (also Teachers Guide). Heinemann.
6 Darlington, C.D & Bradshaw, A.D. (Eds). (1963). *Teaching Genetics in School & University.* Oliver & Boyd.
7 Darlington, C.D & La Cour, L.F. (1969). *The Handling of Chromosomes.* 5th edn. Allen & Unwin.
8 Department of Education and Science (1967). *Safety at School.* Education Pamphlet No 53. HMSO.
9 Department of Education and Science (1973). Safety Series No 2. *Safety in Science Laboratories* HMSO.
10 Educational Use of Living Organisms Project (Schools Council) Kelly, P.J & Wray, J.D. (Eds). (1975). *The Educational Use of Living Organisms. A Source Book.* Hodder and Stoughton Educational.
11 Fisher, R.A. (1936). Has Mendel's work been rediscovered? *Annals of Science* 1 (2).
12 Head, J.J & Dennis, N.R. (1968). *Genetics for O Level.* Students Text (also Teachers Guide) Oliver & Boyd.
13 Nuffield Foundation Science Teaching Project. Advanced Science: Biological Science Fry, P.J. (Ed). (1971). *Laboratory Book.* Penguin.
14 Nuffield Foundation Science Teaching Project. Advanced Science: Biological Science. Gray, J.H. (Ed). (1970). *Organisms and Populations. A Laboratory Guide.* Penguin. (See also 17).
15 Nuffield Foundation Science Teaching Project. Advanced Science: Biological Science. Sands, M.K. (Ed). (1970). *The Developing Organism. A Laboratory Guide.* Penguin. (See also 18).
16 Nuffield Foundation Science Teaching Project. Advanced Science: Biological Science (1970). *Teachers Guide 1 Maintenance of the Organism. Organisms and Populations.* Penguin.
17 Nuffield Foundation Science Teaching Project. Advanced Science: Biological Science (1970). *Teachers Guide 2. The Developing Organism. Control and Co-ordination in Organisms.* Penguin.
18 Nuffield Foundation Science Teaching Project. Biology (1966). Text 1: *Introducing Living Things.* (also *Teachers Guide 1*). Longmans/ Penguin.
19 Nuffield Foundation Science Teaching Project. Biology (1966). Text 2: *Life and Living processes* (also *Teachers Guide 2*). Longmans/ Penguin.
20 Nuffield Foundation Science Teaching Project. Biology (1967). Text 5: *The Perpetuation of Life* (also *Teachers Guide 5*). Longmans/ Penguin.
21 Nuffield Foundation Science Teaching Project. Secondary Science (1971). Theme 2: *Continuity of Life.* Longmans.
22 Parker, R.E. (1973). *Statistics for biology* (Studies in Biology No 43). E. Arnold.
23 Schatz, A., Brandon, G.C & Webber, J.D. (1970). Sterilisation of plastic Petri dishes. *American Biology Teacher* 32 (5), 294-5.
24 Schools Council (1974). *Recommended Practice for Schools Relating to the Use of Living Organisms and Material of Living Origin.* Hodder and Stoughton Educational.

25 Smith, I & Fernberg, J.G. (1965). *Paper and Thin Layer Chromatography and Electrophoresis.* 2nd edn. Shandon.

26 Wallace, M.E., Gibson, J.B & Kelly, P.J. (1968). Teaching genetics: the practical problems of breeding investigations. *Journal of Biological Education* 2 (4), 273-304.

Section 2 **Animals (excluding the Human)** 27-77

27 Balls, M & Godsell, P.M. (1972). Animal cells in culture—methods for use in schools. *Journal of Biological Education* **6** (1) 17-22.

28 Barrass, R. (1974). *The Locust: A Guide for Laboratory Practical Work.* 2nd edn. Barry Shurlock.

29 Burch, A.J.B & Patterson, C.M. (1965). Land snails for demonstrating mitosis and meiosis. *American Biology Teacher* 27 (3), 203-7.

30 Carolina Biological Supply Co./Gerrard & Haig Ltd. (1970). *Exercises with Drosophila.*

31 Crew, F.A.E & Lamy R. (1935). *Genetics of the Budgerigar.* Watmoughs. Idle. Bradford.

32 Demerec, M. (Ed). (1950). *Biology of Drosophila.* Hafner Publishing Co. New York.

33 Demerec, M & Kaufmann, B.P. (1969). *Drosophila Guide.* 8th edn. Carnegie Institution of Washington.

34 Department of Education and Science. (1971). *Keeping Animals in Schools. A Handbook for Teachers.* HMSO.

35 Dyson, H. (1970). *Rabbits.* Cassell.

36 Educational Use of Living Organisms Project (Schools Council). Wray, J.D. (1974). *Small Mammals.* English Universities Press.

37 Flagg, R.O & Noah, L.J. (1970). *Drosophila* mutants. *Carolina Tips* 33 (13). Carolina Biological Supply Co.

38 Green, E.L. (1966). *The Biology of the Laboratory Mouse.* 2nd edn. McGraw Hill. New York.

39 Great Britain, Medical Research Council. (1956). *An Annotated Catalogue of the Mutant genes of the House Mouse.* Gruneberg, H. HMSO.

40 Griffin Biology Experimental Notes (1972). *Keeping Guppies for Genetic Experiments.* Griffin and George Ltd.

41 Griffin Biology Experimental Notes (1972). *Keeping Mice for Genetics* Ala, *Genetic Experiments with Mice* Alb. Griffin and George Ltd.

42 Gruneberg, H. (1952). *The Genetics of the Mouse* 2nd edn. Nijhoff. The Hague.

43 Harris Biological Supplies Ltd (1972). *A Guide to the Maintenance & Use of Locusts in Teaching Laboratories.*

44 Haskell, G. (1961). *Practical Heredity with Drosophila.* Oliver & Boyd.

45 Haskins, C.P & Haskins, E.F. (1948). Albinism, a semi-lethal autosomal mutation in *Lebistes reticulatus. Heredity* 2 (2), 251-62.

46 Haskins, K.P. (No Date) *Using Tribolium for Practical Genetics.* Harris Biological Supplies Ltd.

47 Hoste, R. (1968). The use of *Tribolium* beetles for class practical work in genetics. *Journal of Biological Education* 2 (4), 365-72.

48 Hunter-Jones, P. (1966). *Rearing and Breeding of Locusts in the Laboratory.* Centre for Overseas Pest Research.

49 Hutchings, R.S. (1969). *Fancy Mice.* Cassell.

50 Jacobs, K. (1971). *Livebearing Aquarium Fishes.* Studio Vista.

51 Lynch, G. (1971). *Canaries in Colour.* Blandford.

52 Mertens, T.R. (1971). Additional notes on *Drosophila. Carolina Tips* 34 (8). Carolina Biological Supply Co.

53 Moriarty, F. (1969). The laboratory breeding and embryonic development of *Chorthippus brunneus* Thunberg (Orthoptera: Acrididae). *Proceedings of the Royal Entomological Society of London (A)* 44 (1-3), 25-34.

54 Mylechrest, M. (1969). Timetable of organisation for genetics experiments with *Drosophila. School Science Review* **50** (172), 569-70.

55 Parslow, P. (1969). *Hamsters.* Cassell.

56 Pipino, N.J. (1968). A *Drosophila* mutant to simplify the study of salivary gland chromosomes. *American Biology Teacher* 30 (5). 379-86.

57 Punnett, R.C. (1923). *Heredity in Poultry.* Macmillan.

58 Rogers, C.H. (1964). *Zebra Finches.* Iliffe.

59 Sandford, J.C. (1957). *The Domestic Rabbit.* Crosby Lockwood.
60 Searle, A.G. (1968). *Comparative Genetics of Coat Colours in Mammals.* Logos Press/Academic Press.
61 Shorrocks, B. (1972). *Drosophila.* Ginn.
62 Short, D.J & Woodnott, D.P. (Eds) (1969). *The Institute of Animal Technicians Manual of Laboratory Animal Practice and Techniques.* 2nd edn. Crosby Lockwood.
63 Sokoloff, A. (1966). *The Genetics of Tribolium and Related Species.* Academic Press.
64 Sole, A. (1969). *Cavies.* Cassell.
65 Strickberger, M.W. (1962). *Experiments in Genetics with Drosophila.* Wiley.
66 Taylor, T.G & Warner, C. (1961). *Genetics for Budgerigar Breeders.* Iliffe.
67 Universities Federation for Animal Welfare (1971). *Handbook on the Care and Management of Farm Animals.* Churchill Livingstone.
68 Universities Federation for Animal Welfare (1972). *Hand Book on the Care and Management of Laboratory Animals.* 4th edn. Churchill Livingstone.
69 Uvarov, B.P. (1966). *Locusts and Grasshoppers A handbook for their Study and Control.* Cambridge University Press.
70 Wallace, M.E. (1965). Using mice for teaching genetics I & II. *School Science Review* 46 (160), 646-58 and 47 (161), 39-52; and as *Science Reprints No 1.* Association for Science Education.
71 Wallace, M.E. (1971). *Learning Genetics with Mice.* Heinemann.
72 Watmough, W. (1960). *The Cult of the Budgerigar,* 5th edn. Iliffe.
73 Watmough, W.J. (1969). *Budgerigars.* Cassell.
74 Wilcox, F.A. (1971). Disposable bottles for fruit fly culture. *American Biology Teacher* 33 (9), 544.
75 Williams, R.D & Smith, D.M. (1970), Preparation of blowfly salivary-gland chromosomes. *American Biology Teacher* 32 (8), 491-2.
76 Winge, O. (1927). The location of eighteen genes in *Lebistes reticulatus. Journal of Genetics* 18 (1), 1-43 with three colour plates.
77 Winge, O & Ditlevsen, E. (1947). Colour inheritance & sex determination in *Lebistes. Heredity* 1 (1), 65-83.

Section 3 **The Human Animal** 78-82

78 Clarke, C.A. (1970). *Human Genetics and Medicine.* E. Arnold.
79 Lawler, S.D & Lawler, L.J. (1953). *Human Blood Groups and Inheritance.* 2nd edn. Heinemann.
80 Mertens, T.R. (1970). Human chromosomes. *American Biology Teacher* 32 (8), 468-73.
81 Penrose, L.S. (1961). *Recent Advances in Human Genetics.* Churchill.
82 Stern, C. (1960). *Human Genetics.* 2nd edn. W.H. Freeman.

Section 4 **Flowering plants 83-103**

83 Cameron, J.A. (1970). An introduction to the practical study of variation using *Ranunculus* species. *Journal of Biological Education* 4 (1) 19-24.
84 Clapham, A.R., Tutin, T.G & Warburg, E.F. (1962). *Flora of the British Isles.* 2nd edn. Cambridge University Press.
85 Clevenger, S. (1964). *Flower Pigments.* Scientific American Offprint No 186.
86 Clifford, D. (1970). *Pelargoniums including the popular 'Geranium'.* 2nd edn. Blandford.
87 Darlington, C.D. (1963). *Chromosome Botany and the Origins of Cultivated plants.* 2nd edn. Allen & Unwin.
88 Darlington, C.D & Wylie, A.P. (1955). *Chromosome Atlas of Flowering Plants.* Allen & Unwin.
89 Dayton, T.O. (1956). The inheritance of flower colour pigments. *Journal of Genetics* 52 (2), 249-60.
90 Educational Use of Living Organisms Project (Schools Council). Bingham, C.D. (1976). *Plants.* Hodder and Stoughton Educational.

91 Fuller, J. (1966). Sweet Peas and the School Garden. *The Sweet Pea Annual 1966.* The National Sweet Pea Society.
92 Hutchinson, Sir J. (Ed) (1965). *Essays on Crop Plant Evolution.* Cambridge University Press.
93 Mylechreest, M. (1971). The secondary school garden. *School Science Review* 53 (182), 77-84.
94 Neuffer, G., Jones, L & Zuber, M.S. (1968). *The Mutants of Maize.* Crop Science Society of America, Madison, Wisconsin, USA.
95 Park, B. (1962). *Collins Guide to Roses.* Collins.
96 Phinney, B.O. (1961). Dwarfing genes in *Zea mays* and their relations to gibberellins. in *Plant Growth Regulation.* 489-501. Iowa State University Press.
97 Royal Horticultural Society. (1969). *Dictionary of Gardening.* 2nd edn. & Supplement. Oxford University Press.
98 Satterfield, S.K & Mertens, T.H. (1972). Teaching meiosis with *Rhoeo discolor. Carolina Tips* 35 (3), Carolina Biological Supply Co.
99 Savage, J.R.K. (1967). Demonstrating cell division with *Tradescantia. School Science Review* 48 (166), 771-82.
100 Schiotz, A & Dahlstrom, P. (1972). *Collins Guide to Aquarium Fishes and Plants.* Collins.
101 Shambulingappa, K.G. (1966). A simple leaf-tip squash method for the study of chromosomes. *School Science Review* 47 (162), 487-9.
102 Thomas, G.S. (1963). *The Old Shrub Roses.* 4th edn. Phoenix House.
103 Wheldale, M. (1907). Inheritance of flower colour in *Antirrhinum majus. Proceedings of the Royal Society B* 79 288-305.

Section 5 Micro-organisms, fungi and bacteria 104-113

104 Anderson, G.E. (1971). *The Life History and Genetics of Coprinus lagopus.* Harris Biological Supplies Ltd.
105 Bevan, E.A & Woods, R.A. (1962). Complementation between adenine requiring mutants in yeast. *Heredity* 17 (1), 141.
106 Collins, C.H & Lyne, P.M. (1970). *Microbiological Methods.* 3rd edn. Butterworths.
107 Cunnell, G.J. (1963-1965). Studying fungi in schools. *School Science Review* 4 (155) 93-102 and (156) 312-23; 46 (160) 579-91.
108 Dade, H.A & Gunnell, J. (1969). *Class Work with Fungi.* Commonwealth Mycological Institute.
109 Educational Use of Living Organisms Project (Schools Council). Fry, P.J. (1976). *Micro-organisms.* Hodder and Stoughton Educational.
110 Garbutt, J.W & Bartlett, A.J. (1972). *Experimental Biology with Microorganisms* Students Text (also Teachers Guide). Butterworths.
111 Head, J.J. (1962). Sixth form genetical experiments with *Neurospora crassa. School Science Review* 43 (1950). 437-40.
112 Horn, P & Wilkie, D. (1966). The use of Magdala Red for the detection of auxotrophic mutants of *Saccharomyces cerevisiae. Journal of Bacteriology* 91 (3), 1388.
113 Nicholas, D.J.D & Fielding, A.A. (1951). Use of *Aspergillus niger* for the determination of the magnesium, zinc, copper and molybdenum available in soil to crop plants. *Journal of Horticultural Science* 26 (2), 125-47.